IEE MONOGRAPH SERIES 11

Acoustics for radio and television studios

IEE MONOGRAPH SERIES 11

Acoustics for radio and television studios

Christopher Gilford, M.Sc., Ph.D., F.Inst.P., C.Eng., M.I.E.E.
Reader in Acoustics,
University of Aston in Birmingham,
Birmingham, England

Peter Peregrinus Ltd. 1972
on behalf of the
Institution of Electrical Engineers

Published by Peter Peregrinus Ltd.,
2 Savoy Hill, London WC2R 0BP, England

ISBN : 0 901223 23 9

Printed in England by
Billing & Sons Ltd., Guildford and London

The word 'acoustics' in the title of this monograph covers not only the properties of the studio as an enclosure modifying the sound of a programme originating within it, but also the unwanted background noise from sources inside and outside it and the transmission of sound between the studio and other areas. The study of noise and sound transmission can be pursued by the objective methods common to most scientific inquiry, though there are difficulties caused by the complexity of the hearing organ on the one hand and the number of variables in the transmission process on the other. By contrast, the science of room or auditorium acoustics is, perhaps, unique in the depth of its subjective content. The sense of hearing, though evolved primarily for information gathering and communication, is also intimately associated with aesthetic sensations, emotions and memories. For this reason, the diagnostic aspects of studio acoustics—the search for parameters correlated with aesthetic satisfaction—will inevitably remain the most intriguing and baffling part of the subject.

The powers of the brain for storage, analysis and pattern recognition make the most sophisticated efforts of modern science appear crude and immature, and it is not surprising, therefore, that the simple parameters that have been evolved in an attempt to distinguish between good acoustics and bad have proved largely inadequate. They do, indeed, help one to design and criticise particular studios, but there is yet no unifying theory that can be used as a general guide.

One must not, however, lapse into despair or obscurantism. A great deal has been achieved in the recent past in improving the standard of studio design, as one may easily discover by comparing speech recordings of, say, 20 years ago with the present daily output of broadcast speech, and there is no reason to suppose that there will not be further advances. This monograph is an attempt to summarise the results of research in the acoustics of studios since the end of the Second World War.

Prewar broadcasting research had established the importance

of accurate control of studio acoustics and of the suppression of noise. Minimum studio sizes had been agreed and the techniques of measurement of reverberation times and absorption coefficient had been developed. Studios tended to be excessively reverberant at low frequencies since satisfactory resonant absorbers had not been devised, and there was a need for a range of selective absorbers with precisely predictable characteristics.

The methods of radar scanning and improved pulse circuits held promise of fruitful application to studio acoustic diagnosis, in association with the theoretical work on wave acoustics that had been done in the USA during the previous ten years. The dramatic reductions in the sizes of valves and other circuit components made it possible to build light, easily portable apparatus, not only for carrying out existing acoustical measurements but also for tests and measurements based on new and promising techniques requiring more complicated circuits. Such apparatus could be carried around from one studio to another, to make measurements by which new hypotheses could be tested and statistical data accumulated. A further stimulus came from the rapid improvement of magnetic recording, which greatly facilitated subjective tests.

In the following years, the acoustics section of the research department of the British Broadcasting Corporation probed every aspect of acoustics and noise control in studios, developing diagnostic tests, carrying out acoustic measurements and experimental modifications of operational and prototype studios, and assembling specifications for design. In doing so, we were continually aware, through the literature and personal contacts, of current work and ideas here and abroad; the present review, though it may in some respects appear to be a personal statement, has been deeply influenced by these contacts.

C. L. S. GILFORD

ACKNOWLEDGMENTS

The author wishes to thank the Director of Engineering, BBC, for permission to use much of the material in this book, and acknowledges with the greatest gratitude the contributions made over the years by past and present members of the BBC research department, particularly T. Somerville, formerly head of the electroacoustics group, and an outstanding pioneer in the subject. Thanks are also due to W. Kuhl of the IRT, Hamburg, for many stimulating discussions and letters on matters of common interest over the years.

The author wishes to acknowledge permission to use certain Figures and graphs from other sources. Extracts from BS4198:1967 'Method for calculating loudness', BS3639:1963 'Measurement of sound absorption coefficients', and BS3383:1961 'Equal loudness contours', are reproduced by permission of the British Standards Institution, 2 Park Street, London W1A 2BS, from which copies of the complete standards may be obtained. Figs. 1.1a and 8.6 are reproduced by permission of the Institute of Physics; Figs. 1.2, 1.3 and 3.2d of Her Majesty's Stationery Office; Figs. 1.1b, 1.7, 1.8, 1.9, 2.6, 3.2d and 9.8 of the Editors of *Acustica*; and Figs. 3.4 and 4.6 of the Acoustical Society of America. All the plates are reproduced by courtesy of the BBC, as are also Figs. 3.1, 3.2a, 4.2, 4.8, 4.20, 5.6, 5.11, 6.5, 6.6, 7.13, 8.3, 8.7–8.10, 8.13–8.17, 9.3, 9.4, 9.6, 9.7, 10.4, 10.5, 11.1, 11.4, 11.6, 12.1 and 12.3. The authors concerned are acknowledged in the legends or in the main text as appropriate, and thanks for permission are due also to them.

Contents

A*

PLATES

Sound and hearing

1.1 Introduction

In this chapter, it is proposed to give only the definitions essential for understanding the text, together with a description of the ear and of the very remarkable functions that it performs, in so far as they affect the requirements and design of broadcasting studios. It is assumed that the reader has access to standard modern textbooks and that he can, therefore, make himself fully acquainted with the main properties of wave motion, vibration and sound and with the related experimental and theoretical techniques that have been developed.

Appendix 13.1 contains a bibliography of such texts, and covers aspects of acoustic theory and practice that will be found helpful. The list has been kept as short as possible, and the books representing each branch of the subject have been chosen for readability and simplicity of treatment as much as for other qualities. References to books and papers are also included in the usual manner in the text where required to illustrate specific points.

1.2 Nature of sound

1.2.1 Elementary concepts and definitions

Sound is the name given to the movements of the particles of a solid or fluid which are propagated through the medium by reason of its elasticity. The name is also given to the sensation created when the moving particles impinge on the ear drum; the word will here be used to indicate only those particle movements which are of such a nature as to be heard by a normal human ear.

Sound is generated in a medium by any process that causes a local movement in any part of the medium. The movement may be

impulsive, consisting of a sudden displacement lasting only a short time, or it may be **periodic**, consisting of a pattern of displacements indefinitely repeated. With the latter, the number of such repetitions in unit time is known as the **frequency** of the sound, and this frequency largely determines the sensation of **musical pitch** that is associated with such periodic sounds.

The normal limits of frequency of audible sound are from about 30 Hz to 20 kHz. The upper limit of hearing depends on the age of the hearer, falling steadily with age after about 20 years.

The displacements in a medium may be neither finite in duration nor periodic, but instead consist of a time-sequence with no discernible structure; sound resulting from this type of displacement is called **random sound**. **Noise** is usually defined as sound that is undesired by the recipient: it is treated in this sense in Chapter 3 and elsewhere. However, the term is frequently used by the electrical engineer to denote a deliberately created random signal that contains a wide range of frequencies, which is of value in investigating transmission systems.

The **spectrum** of a sound is the relationship, usually shown in the form of a graph, of the power of the sound energy in unit bandwidth with the frequency. The **time function** of a sound and its spectrum are related by integral equations associated with the name of Fourier (see Appendix 13.1). Thus a continuous sinusoidal signal possesses a spectrum consisting of a single frequency; conversely, the spectrum of a single-impulsive sound is a continuous function containing all frequencies.

The **instantaneous particle velocity** in a sound wave is the velocity of a particle of the medium due to the propagation of the sound. The effective value at a point is generally expressed as the square root of the mean value of the square of the velocity (the **root-mean-square, or r.m.s., velocity**) averaged over a complete cycle of the movement, or, for a nonperiodic sound, over a sufficiently long time.

Adjacent particles do not, in general, move simultaneously, those further from the source of the sound repeating the motion carried out some time previously by those nearer the source. Compressions and rarefactions therefore occur at every point in the medium, and the **instantaneous sound pressure** is the fluctuating sound pressure in the sound wave associated with these compressions and rarefactions. Its effective value at any point is the root-mean-square average over the period of the oscillation, or over a long time for a nonperiodic sound.

The **intensity** of sound at a point in the medium is the rate at which sound energy flows through a surface in the medium perpendicular to the direction of propagation and of unit area. If p is the r.m.s. pressure, the intensity is given by $p^2/\rho_0 c^2$ (W/m^2) where ρ is the density of the air at rest and c is the velocity of sound in the medium.

Decibel notation: In comparing loudnesses of two sounds, the ear takes notice of the **ratios** of their intensities rather than their **differences**, and it is thus convenient to express the intensity of a sound wave in terms of its logarithm instead of the intensity itself. The particle velocity and the pressure are related to the intensity by a power law (intensity $\propto$ pressure^2 or $\propto$ velocity^2), and therefore the same applies to these quantities also.

Thus

$$\text{the Intensity level} = 10 \log(I/I_{ref}) \text{ decibels}$$

where I is the intensity and I_{ref} is a specified reference intensity, actually 10^{-12} W/m^2 by international agreement.

Similarly, the **sound-pressure level** (s.p.l.) is

$$20 \log_{10}(p/p_{ref})$$

where $p_{ref} = 2 \cdot 10^{-5}$ N/m^2.

For plane waves in air, these two expressions are almost exactly equal, the intensity level of a given sound being approximately $0 \cdot 5$ dB lower than the pressure level as defined above. (For other media, the expressions will generally differ more widely.) It is therefore permissible to regard the intensity level and the pressure level of a sound in air as equal for most purposes, but it must be borne in mind that the equality is largely coincidental.

Characteristic impedance: By analogy with electrical-circuit theory (Olson and Massa, 1943), pressure may be considered as potential difference and velocity as current. The ratio of the two in free plane-wave propagation is known as the characteristic impedance of the medium, and is equal to ρc, where ρ is the density of the medium and c is the velocity of sound in it. (Other analogies are possible, the most useful alternative being the 'mobility' analogy, in which velocity corresponds to voltage and pressure to current.)

Acoustic impedance: The acoustic impedance at a given surface is the ratio of the pressure at the surface to the volume flow through it (i.e. the rate of flow of the medium perpendicularly through the surface, or the product of the velocity and the area). Note that this refers to a specific surface, and not to unit area. The acoustic impedance is generally a complex quantity.

Mechanical impedance: The acoustic impedance defined above is useful for the examination and analysis of acoustic-transmission paths. The mechanical impedance is defined as the ratio of a force to a velocity. It is usually applied to mechanical systems, but it is also useful in the analysis of transducers or sound absorbers of finite area (see Chapter 8).

The **strength of a source** of sound is the total volume flow (r.m.s.) at the surface of the source or through a surface totally enclosing it.

1.2.2 Limits of sensation

Frequency: As already mentioned, the ear is sensitive to airwaves of frequencies from about 30 Hz up to 20 kHz. The upper limit is partly determined by losses in the transmission of the sound from the ear drum to the sensory elements in the inner ear. Sound of too high a frequency to be audible is called **ultrasound** (or **ultrasonic**), and sound of too low a frequency to hear is called **infrasonic.** By bone conduction, it is possible to hear much higher frequencies than through the ear; e.g. a transducer vibrating at 50 kHz can be

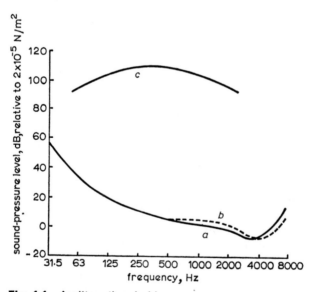

Fig. 1.1 Auditory thresholds

  a Mean threshold of hearing, pure tones, progressive waves (BS3383:1961)
  b Mean threshold of hearing, octave bands, diffuse field (Robinson and Whittle, 1964)
  c Threshold of pain

heard by placing it against the cheekbones, though there is little definite sensation of pitch.

Vibrations below the lower frequency limit of audition can be felt by any part of the body, but there is no sensation of hearing.

Amplitude: The average threshold of audibility at 1 kHz is at a pressure of $10^{-5} \, \text{N/m}^2$. Fig. 1.1, curve *a*, shows the threshold pressure level as a function of frequency for an average subject listening to free progressive waves. Curve *b*, the threshold for narrow bands of noise in a diffuse field, in which the sound reaches the subject equally from all directions, is rather more useful to the studio designer since it refers to more typical conditions of hearing.

Threshold of pain: Very high sound pressures produce pain in the ears, the threshold pressure levels being shown in curve *c* of Fig. 1.1. At such levels, comparatively short periods of exposure to the sound will cause shifts in the threshold of hearing that although in the first place temporary, may become permanent.

1.3 Structure of ear

1.3.1 Evolution

The ear has evolved to fulfil its specialised purpose in a most remarkable manner, and has done so presumably through the normal processes of mutation and natural selection (Stevens *et al.*, 1965), by reason of the advantages conferred on a species by the possession of an efficient hearing apparatus that enabled them to avoid predators and to find prey. Evolution of the ear probably started in our aquatic ancestors, from an organ similar to the swim bladder in fishes, which is used for regulating their buoyancy. This organ would already be sensitive to fluctuations in pressure, and, under water, it would be in a favourable environment, because energy would be transmitted into the tissues without large losses due to reflection. Evolution would have been in the direction of progressive refinement of the ear and the associated parts of the central nervous system, so as to achieve the extraordinary sensitivity, dynamic range, and discrimination of frequency, quality and direction, with which the ear is endowed. The elegant and subtle methods by which these properties are achieved are described in this Section. It is also hoped to provide an insight into the ability of the listener to recognise, in a manner that is quite beyond the practicable capabilities of computers, the properties of a sound as a pattern or *gestalt*.

1.3.2   Physical structure

The structure and function of the ear has been explored in great detail by many workers, and it is possible here to give only a very brief description. It is hoped that the form of this description will appeal to the electrical engineer. For a fuller account, the reader is referred to a very readable paper by Burns (1962).

Fig. 1.2 is a functional diagram of the human ear. The pinna surrounds the entrance to the ear canal, which is terminated at its inner end by the ear drum. The pinna has two possible uses: being conical in shape, it may perform a slight impedance-matching function, and it also probably assists in the determination of the direction of a source of sound (Tarnoczy, 1958). The length of the ear canal averages 25 mm, and the diameter 7 mm. This length would give a resonance frequency of 3·3 kHz, which corresponds approximately to the frequency of maximum sensitivity of the ear, though there is some disagreement about the relationship.

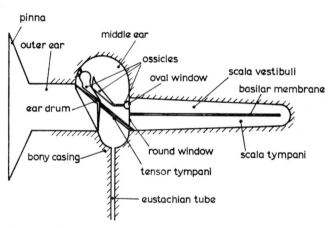

Fig. 1.2   Functional diagram of human ear

The ear drum is conical in shape with an oval base, the vertex being inwards. It forms the outer boundary of the middle ear, which is a closed cavity filled with air and is connected to the nasal passages through the Eustachian tube. This tube is opened momentarily by the act of swallowing, allowing the pressures inside and out to become equal. The tensions due to unequalised pressures and the relief obtained by swallowing will be familiar to any one

who has travelled by air. On the opposite side of the middle ear
to the ear drum are two membranes, the oval window and the
round window, which form seals between the middle ear and the
complex mechanism of the inner ear. Movements of the ear drum
caused by external pressures are transmitted to the oval window
by a chain of small bones forming a lever system with a high mechanical advantage. This chain forms a mechanical impedance transformer, which reduces the amplitude of the motion to a value more
suitable for transmission through the fluids contained in the inner
ear. The transformer ratio can be varied over a range of about
10:1 to provide protection for the inner ear should the ear drum
be deflected to a dangerous extent. This variation is carried out
by alterations in tension of two muscles, the tensor tympani,
attached between the ear drum and the inside of the cavity, and the
stapedius, which extends from the endbone to the side wall.

The inner ear, which is completely encased in bone, has the form
of a fluid-filled spiral cavity, called the cochlea from its resemblance
to a sea shell. In Fig. 1.2, the spiral is shown as if unwound, and
Fig. 1.3 is a simplified diagram of the cross-section of a turn of the

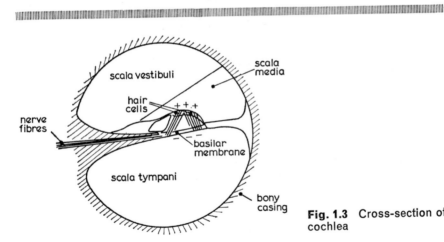

Fig. 1.3 Cross-section of cochlea

cochlea. The cochlea is separated longitudinally into three passages:
the larger two are the scala vestibuli, leading from the oval window,
and the scala tympani, ending in the round window. These two
passages communicate at the far (small) end of the cochlea. The
division between the lengths of these two passages is formed by a
bony shelf, which extends almost across the section, and by the

basilar membrane, which completes it. The basilar membrane increases in width from about 0·1mm at the end nearest to the oval window to 0·5mm at the far end, its length being about 31mm. The mechanical properties of this system are such that a sound wave transmitted along it from the oval window causes a deflection of the basilar membrane, which increases at first and then diminishes, the point along the membrane at which it attains maximum amplitude depending on the frequency of the sound. Maximum amplitude is reached near the window for the highest, and at the far end for the lowest frequencies. Fig. 1.4 illustrates this process. Supported along the basilar membrane is the organ of corti, which consists of an assemblage of terminal nerve elements ending in fine-hair cells, which project from it and end in a parallel membrane called the tectorial membrane, which is hinged on the bony shelf. Movements of the basilar membrane and the tectorial membrane, which occur in phase, cause the hair cells to be sheared or bent.

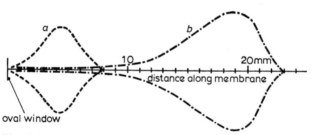

Fig. 1.4 Variation of excitation along basilar membrane
a High-frequency tone
b Low-frequency tone

The organ of corti and the tectorial membrane are enclosed in an inner duct called the scala media, which is separated from the main part of the scala vestibuli by Reissner's membrane. This membrane serves as an electrically insulating septum between the fluid in the main passages on the one hand and that surrounding the nerve cells, which is maintained at a slightly different potential. There is thus a small, steady leakage current through the nerve cells, which is modulated by the shearing action to which they are subject, giving rise to measurable potentials proportional to the sound pressures on the ear. These are called the cochlear microphonic potentials.

1.3.3   Auditory nerve system

The cochlear microphonic potentials, acting in a manner that is not fully understood, produce repeated electrical impulses, which travel through the nerve pathways to the brain. These impulses last about 1 ms, during which time the nerve cannot be restimulated, about 20 ms elapsing before the nerve is restored to its full action. The number of impulses is a function of the intensity of the stimulus, and thus the location of the nerve that transmits the greatest number of impulses determines the frequency that is sensed. Galambos and Davis (1943) obtained an increase from zero to about 400 impulses/s with an increase of stimulus of 25 dB.

The details of the auditory pathway are of bewildering complexity, and cannot be said to be fully understood. Present knowledge and speculation have been admirably summarised by Whitfield (1967). Only the resulting achievements of the system will be described here.

1.4   Ear as sound analyser

1.4.1   Sensitivity

The curve of sensitivity against frequency has been given in Fig. 1.1. The minimum audible sound pressure for an ear of exceptional sensitivity is about $8 \times 10^{-6} \mathrm{N/m^2}$, corresponding to a displacement amplitude of $1 \cdot 25 \times 10^{-12} \mathrm{m}$ in air. This is only one-hundredth of the diameter of a hydrogen molecule, and considerably less than the Brownian movement in the ear cells. This astonishing sensitivity must make use of some form of correlation mechanism in the central nervous system capable of detecting minute information-bearing signals some 40–60 dB below noise.

1.4.2   Loudness of sound

The loudness of a sound can be defined as the magnitude of the sensation that it produces. This is measured most conveniently by finding the pressure level of a 1 kHz tone that is judged equally loud. The s.p.l. of the comparison tone is then known as the loudness level of the sound, and is expressed in phons. The loudness level in phons can be converted into loudness in a subjective scale known as the sone scale, in which doubling the number of sones corresponds to a doubling of the subjective loudness of the sound.

An advantage of the sone scale is that loudnesses in sones may be added to give an aggregate loudness in sones that may be converted into loudness level in phons. This process may be carried out to obtain the loudness or the loudness level of a complex sound containing a wide range of frequencies. Before we can do this, however, it is necessary to understand the masking of one part of the audible frequency range by another. It will be remembered that, if two tones are received by the ear, two parts of the basilar membrane will be excited, and it is only if these two regions do not overlap that the loudnesses can be summed independently. Similarly, if a narrow band of noise of constant power but increasing width is applied to the ear, the loudness will be independent of the bandwidth until it exceeds a certain critical amount, when the loudness will start to increase because the ends of the band are far enough away from each other not to mask each other. This bandwidth is known as the critical bandwidth, and, for purposes of summation, all components of a sound lying within a critical bandwidth (approximately ⅓ octave over most of the audible frequency range) behave as if they were a single tone of the central frequency and of the same power.

Methods of calculating the loudness of a complex sound taking masking into consideration are described in Section 2.4.

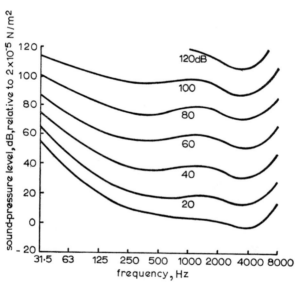

Fig. 1.5 Curves of equal loudness for pure tones in progressive wave field (after BS3383:1961)

1.4.3 Sensitivity to differences in loudness

The discrimination between nearly equal loudnesses for pure tones follows the general shape of the equal-loudness contours shown in Fig. 1.5. The greatest discrimination occurs at about 4kHz, and increases with increase of pressure level. At 4kHz, a change of 3dB is just noticeable in a tone 5dB above threshold, but a change of 0·5dB can be heard at 80dB above threshold (Riesz, 1928). At 80Hz, the figures are approximately 6 and 0·6dB, respectively. Other workers have obtained figures of a similar order, though detailed results vary considerably.

A more important figure for the broadcasting engineer is the just-noticeable difference in programme level under good experimental conditions. This is about 1·2dB for both speech and music; a stepped volume control with 1dB increments therefore gives a sufficient illusion of continuous variation in loudness without noticeable discontinuities.

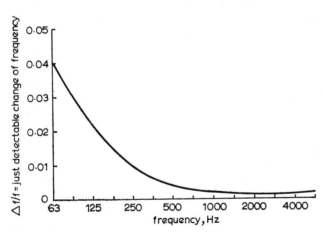

Fig. 1.6 Frequency discrimination of average ear (after Knudsen, 1923)

1.4.4 Frequency discrimination

The ear shows a very remarkable power of discrimination of pitch or frequency. Fig. 1.6 shows the results of Knudsen (1923), in which the ratio of the minimum perceptible frequency change to the absolute frequency $\Delta f/f$ is plotted against the frequency at a loudness level of 60 phon. The ratio increases slowly at all frequencies as the loudness level is decreased.

The minimum perceptible change is least at 2 kHz, with an average value of 0·0017; unusually sensitive or trained ears can detect changes as small as 0·001, which corresponds to a length on the basilar membrane of less than 0·1 mm. What is, possibly, more remarkable is that many musicians, from their familiarity with standard musical sounds, can tune a stringed instrument within this order of accuracy without the necessity to refer to a fixed pitch. The possession of absolute pitch of such accuracy implies a stability in the behaviour of the cochlear mechanism that is hard to reconcile with its structure.

The fine frequency discrimination, moreover, seems surprising in view of the flatness of the curves in Fig. 1.4. Undoubtedly, the exact point of maximum excitation on the basilar membrane must be determined by some process of counting and comparison of the total numbers of pulses coming from the nerve cells in an appreciable period of time. Much the same process is involved in the location of a point of pressure or vibration on other parts of the body. Von Bekesy (1955) made a model of the basilar membrane consisting of a hollow tube about 50 mm in diameter, with a slot, cut parallel to the axis of the tube, varying uniformly in width from one end to the other. One end of the tube was closed by a bellows attached to an electro-magnetic vibrator driven by an audiofrequency oscillator; the other end and the slot were sealed with membranes made of a plastic material. The membrane along the slot was surmounted by a narrow rib of the same material on which the forearm could be laid. He showed that, with increase of frequency, the point of maximum amplitude of the ridge moved steadily along the tube towards the vibrator. Moreover, using the forearm, as described above, as a sensor of the maximum-amplitude point on the rib, this position could be determined very accurately even if the arm was clothed, although, as with the basilar membrane itself, the curve of deflection is very flat.

Assuming that this positional discrimination results from a comparison of the total pulse count from neighbouring nerve cells over an appreciable time, it is possible to obtain some estimate of the length of time necessary for the process. Gilford and Somerville (1950) compared the standard deviation of estimates of frequency in short pulses of tone consisting of 3 or 6 complete wavelengths as judged by a crew of musically aware subjects. The standard deviations were found to be greater in the 3-cycle than the 6-cycle trains, both being greater than the effective bandwidths of the trains as calculated by Fourier analysis (see Appendix 13.1).

If the wavetrains were played first into a reverberant room and received by a microphone before presentation to the subject, it was found that the standard deviation was significantly diminished. Since the spectrum of a number of randomly phased repetitions of a wavetrain is substantially the same as that of a single component train, it must be concluded that the increased precision of pitch judgment was made possible by the additional time for pitch assessment afforded by the reverberation. The reverberation-times used were 1–2 s, enough to cover many hundreds of pulses, using the estimate of Galambos and Davis (1943).

Up to this point the words 'pitch' and 'frequency' have been used as if interchangeable. The subjective sensation of pitch does not, however, depend only on the frequency. If the frequency of the note is doubled, the pitch is raised accurately by an octave, but the sensed pitch depends also on the loudness of the note. On the average, a sound below the frequency of greatest sensitivity appears to fall in pitch as the loudness level rises, and those above that frequency appear to rise.

Subjects differ greatly from one another in the magnitude, and even the sense, of these changes, some experiencing no pitch change even over a level change of 80 dB. On the other hand, the author is very conscious of pitch changes when listening to orchestral music with a large dynamic range.

Many people hear slightly different pitches in their two ears if exposed in turn to a tone of constant pitch (diplacusis).

1.4.5 Spectrum and pattern recognition

The ability of the ear to recognise the characteristics of a sound, and thus identify the probable source, is truly extraordinary, and may be claimed to surpass by far the most sophisticated system yet devised by science. One example will suffice. Most people can recognise hundreds of acquaintances from the sound of their voices alone, yet no speech-synthesis machine has yet succeeded in producing sounds (by combination of parameters derived from the analysis of voices) that are more than just anonymously human. Moreover, we can still identify these same acquaintances at the other end of the telephone, in spite of the inevitable distortions and limitations of spectrum; or, on hearing a strange voice under such circumstances, we may judge the speaker's age and geographical region with considerable accuracy.

This ability to identify the gestalt, or overall characteristic, of a complex auditory time function is a tremendous challenge to acoustics research. It is perhaps true to say that no real advance can be made in solving the problems of aesthetic design in acoustics until this characteristic of the human being is fully understood. Every point in a concert hall will receive a totally different sound signal from a musician on the stage, and yet an experienced listener can identify unerringly a hall with which he is familiar, even through the medium of a monophonic recording.

It is not surprising, therefore, that no theory has yet been developed that will provide an objective appraisal of the acoustics of studios that correlates with a simple good–bad scale of subjective appreciation, though various attempts have been made. One such attempt was that of Somerville (1953), who obtained an empirical combination of parameters that, for large studios and concert halls, gave a significant correlation with subjective goodness.

1.4.6 Recognition of first sound

Haas (1951) found that a reflection or repetition of a sound reaching a listener was heard as a separate sound only if it was more than 50 ms behind the direct sound. If the delay was 5–35 ms, the second sound could not be detected at all, unless it was considerably louder (by more than 10 dB) than the original sound. Between 35 and 50 ms, it was heard as a separate sound, but appeared to come from the direction of the original source, irrespective of their relative directions. Fig. 1.7 is a curve relating the excess level required by an echo for audibility with the time delay.

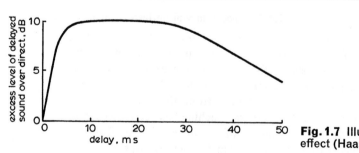

Fig. 1.7 Illustrating Haas effect (Haas, 1951)

This so-called Haas or precedence effect may be caused by the suppression of the nerve potentials immediately following the receipt of a sound; it enables one to identify the source of a sound from the direct wavefront, unaffected by reflections from nearby objects or the surfaces of an enclosure, and is therefore a faculty of evolutionary value. Sound arriving within 50 ms of the direct sound is generally regarded as part of the direct sound instead of the reverberant sound in a studio, and several objective diagnostic tests depend on this grouping (e.g. Thiele, 1953).

Haas's experiments were confined to identical loudspeaker sources in the same horizontal plane. If we have two such loudspeakers and means for varying their relative level, it will be found that the louder of the two appears to be the only source. If the levels are equal, there appears to be a single source halfway between them, the virtual image moving rapidly from one source to the other for slight relative changes about equality of level. For sources in the same vertical plane, the effect is very much less marked, and the combined source starts to move away from the louder of the two even when the level of the weaker source is 15 dB lower.

Closely connected with this property of the ear is the ability to recognise the presence of disturbing echoes. This has been investigated by Muncey, Nickson and Dubout (1953) using speech, string music and organ music, each of fast and slow tempi. Here we enter the province of aesthetic or qualitative judgments, because it is necessary to recognise an impairment in a programme. The reactions of different subjects, therefore, differ much more widely

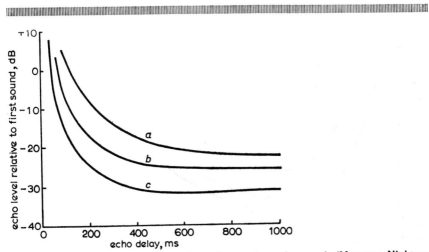

Fig. 1.8 Audibility of discrete echoes in music and speech (Muncey, Nickson and Dubout, 1953)

a Organ music
b String music
c Speech

than with quantitative judgments such as the relative loudness of two sounds, and the authors present the results in the form of curves relating the properties of the echoes with the percentage numbers of subjects finding them disturbing. Fig. 1.8 summarises the relationship between percentage disturbances and the levels and delay times of the echoes, for the three types of programme. Dubout (1958) showed that the curves in this figure can be derived from the assumption that any echo which momentarily exceeds the direct compoment of the sound will be audible. Furduev (1965) drew similar conclusions by studying the autocorrelation functions of programme signals of various kinds.

1.4.7   Discrimination of reverberation time

The discrimination of small differences in reverberation time is of some interest, since it determines the accuracy with which predictions of studio acoustics must be carried out at the design stage to give satisfactory results. The significance and methods of calculation of the reverberation time of a studio are discussed in Chapter 7. Seraphim (1958) presented exponentially decaying pulses of random noise in octave bandwidths to 526 subjects, the pulses being presented in pairs consisting of a test pulse and a standard pulse for comparison. The test and standard pulses differed only in their

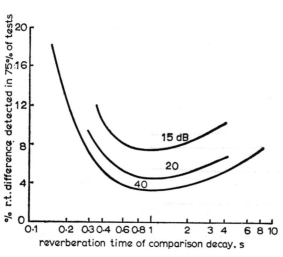

Fig. 1.9   Discrimination of reverberation-time changes (Seraphim, 1958)
      Parameter: signal/noise ratio

times of decay, the differences being up to 18%. The subjects were asked to identify the longer of the decay times, and the results are shown in Fig. 1.9. The three curves represent the differences that were just detectable by 75% of subjects, plotted against the decay time of the standard comparison signal for three different signal/noise ratios. The starting pressure level of each pulse was 80dB relative to $2 \times 10^{-5} \, N/m^3$.

It will be seen that the discrimination is slightly better than 4% in the normally applicable range of reverberation times for studios (0·4–2s) with a signal/noise ratio of 40dB. Further increases of a signal/noise ratio effect little improvement in discrimination, but the least detectable differences increase with increasing rapidity as the background noise increases.

These experiments were carried out in a free-field room, and the pulses constituted a somewhat artificial programme. The minimum detectable change in studio-reverberation time, when judged under normal conditions, is generally accepted to be rather less than 10%. Uniform deviations of this order from a prescribed optimum figure are not likely to be noticed. Changes affecting the two ends of the frequency scale in different directions will, however, be noticeable even if the divergence is everywhere less than 10%.

1.4.8 Recognition of direction of source of sound

The most important factor in determining the position of a sound source is its direction, in the horizontal and vertical planes. It is a matter of common observation that the horizontal direction of a source can be estimated with considerable accuracy, while the vertical direction is much more difficult to judge with certainty. The judgment of horizontal direction undoubtedly makes use of binaural hearing, i.e. the differences between the sounds reaching the two ears. If the source of sound is to the right of the median plane between the ears, the right ear will receive the sound before the left ear and will generally hear it more loudly. Both these differences are important. In general, it is very easy to distinguish right from left; the accuracy is greatest when the sound is directly in front of the observer and least when it is at the side. Complex sounds are easier to locate than pure tones. Mills (1958) finds that the minimum audible angular shift has a minimum value of 1° at 1 kHz.

The exact mechanism of binaural location has been the subject of many investigations from Rayleigh (1876) onwards. A good account is that by Sayers and Cherry (1957).

Recognition of direction in a vertical plane is found very much

B

easier by some individuals than others, and requires a complex, preferably intermittent, sound (Somerville *et al.*, 1966). This faculty is harder to explain than horizontal-direction finding, because sound from a source anywhere in the medial plane of the head may be expected to reach the two ears simultaneously and with equal loudness. Cherry (1963) has stated that small, possibly unconscious, movements of the head are necessary and sufficient, though other observers have found that a subject with his head firmly clamped retains a vertical directional sense. Tarnoczy (1958) has suggested that the distortion of the sound field by the pinnae is sensed by the subject and interpreted in terms of vertical direction. He showed that, if a subject hears through tubes a few centimetres in length inserted into his ears, he loses all vertical directional sense. Thurlow, Mangels and Runge (1967) found that head movements did not materially affect vertical localisation at low frequencies. Roffler and Butler (1968) found that reliable vertical localisation was possible only if the sound was complex and contained frequencies above 7kHz, and if the pinnae were present.

It is possible to judge the distance of a source of sound as well as its direction. Such judgment is probably dependent on the relative levels of the direct and reverberant sounds, but tends to be unreliable. Coleman (1962) found that this faculty improved markedly with practice.

1.5 Conclusion

The information presented in this chapter gives some idea, it is to be hoped, of the extraordinary powers of analysis and pattern recognition possessed by the combination of the two ears and the brain. These powers are based on almost unbelievable sensitivity and discrimination, together with mental data-processing facilities which transcend anything possible to the systems engineer. The following chapters will attempt to show what impairments of a programme, due to unwanted noise distortion or reverberation, can be tolerated, and the positive measures necessary to create an acoustic environment that will give satisfaction. It will be no surprise to find that some of these measures are still partly empirical in nature, depending on experiment and experience rather than reference to any clear basic principles. The ear must always be the ultimate judge of quality, and all methods of measurement, acoustic theories and methods of design must ultimately be tested against its judgments.

Chapter 2

Measurement of noise

2.1 Definitions

The following definitions have particular application to the measurement of noise, and supplement those given in Section 1.2.

Band-pressure level: the sound-pressure level of the sound energy within a specified frequency band (in noise measurement, the frequency bands are normally 1 octave or ½ octave wide; critical bandwidths, mentioned in Section 1.4.2, or approximations to them are also used).

Spectrum-pressure level: the band-pressure level for a bandwidth of 1 Hz centred at a specific frequency

Sound level: The sound-pressure level as measured after a known relative weighting has been applied to the constituent frequency bands; the type of weighting must be specified.

2.2 Measurement of sound-pressure level and band-pressure levels

2.2.1 Basic method of measurement

The essential components of equipment for measuring noise are an accurately calibrated microphone, a microphone amplifier with low inherent noise, bandpass filters and an indicating instrument. A measurement chain giving the utmost flexibility for all kinds of applications is shown in Fig. 2.1.

The microphone is any good omnidirectional type of which the sensitivity and frequency characteristic are known. If it is to be used for the measurement of general background noise in studios as well as noise from specific sources, the mean polar response should also be known, and it may be a good idea to build an equaliser to secure as flat a characteristic as possible, thus avoiding the necessity of applying a correction to every measurement. The sensitivity should be as high as possible in relation to the inherent noise level of the microphone so that measurements may be made of background noise the level of which is comparable with the lowest programme-sound levels occurring in a broadcast.

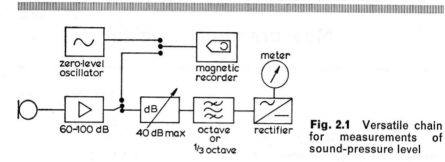

Fig. 2.1 Versatile chain for measurements of sound-pressure level

A suitable gain for the microphone amplifier is 100dB, with input-stage noise close to the thermal noise associated with the input resistance, and with its gain variable in calibrated steps of 10dB.

The calibrated attenuator, which is connected between the output of the amplifier and the input of the bandpass filter, is adjusted to give a fixed indication on the meter, irrespective of the microphone signal; the sound at the microphone is then calculated by simply adding a constant to the attenuator setting in decibels. This constant embodies the sensitivity of the microphone, expressed as the r.m.s. sound-pressure level required to produce 1 mw of power into a $600\,\Omega$ load, the gain of the amplifying chain with the attenuator set to zero and the constant reference level at the indicator.

The indicator will take the form of a rectifier and meter, or a highspeed level recorder. The rectification in either case may be designed to indicate the peak value of the waveform, the mean or the root-mean-square. All single-purpose instruments will agree in reading a pure sinusoidal waveform, since calibration will normally be effected on a known voltage, but an instrument having alternative forms of rectification will show that the peak value of a sinusoid is $\sqrt{2}$ times the r.m.s. (3dB higher), and the mean value is 0·9dB lower than the r.m.s. For other waveforms, the ratio of peak to r.m.s. values will vary widely, reaching as much as 10dB or even more for waveforms of high harmonic content. Moreover, the 'peak' reading instruments used by many broadcasting organisations for monitoring programmes do not indicate the true crest value of a waveform but show instead a value intermediate between r.m.s. and peak.

The following Table shows the differences that can occur between indicating instruments with different types of rectification when measuring the same noise. The figures in the Table represent the amounts by which the measured band-pressure levels of two types of noise exceed those shown by a meter with true r.m.s. characteristics.

Table 2.1   Comparison of noise measurements with various forms of rectification (Gilford and Jones, 1967)

	Difference, dB					
Rectification	Thyratron noise			Ventilation noise		
	Whole	Octave	⅓octave	Whole	Octave	⅓octave
Mean	−0·6	−1	−0·5	0	−0·5	0
Quasipeak (programme meter)	4·2	5	5	5	5	6
True-peak chart recorder	8	4–8	4–8		6–13	3–8

BS3489:1962 makes the use of an r.m.s. meter obligatory for noise measurement, and it is therefore advisable to use this type of rectification unless there are strong reasons for doing otherwise. A simple linear system of rectification giving the mean rectified voltage is a fairly good substitute if an r.m.s. meter is not available. This is illustrated by Table 2.1, and may be taken to apply to most of the noises with which the broadcasting engineer is concerned.

There are occasions, however, on which a peak- or quasi-peak indication is more suitable than the standard r.m.s., and, for a wide range of 'random' noises, the reading given by such indication will be about 5dB above the r.m.s. reading. Further reference to this will be made in Chapter 3.

The filters should be octave- or ⅓ octave-bandpass filters conforming to the requirements of BS2475:1964.

2.2.2 Specialised noise-measuring equipment

The measurement chain described above can be assembled from apparatus available at any broadcasting station, with the possible exception of bandpass filters with the correct characteristics. A single amplifier with 100dB gain may be replaced by low-noise broadcast microphone amplifiers in cascade. There is, unfortunately, a dearth of good commercially available low-noise modular equipment from which a really satisfactory combination may be selected, but it is possible to obtain complete equipment designed for the purpose. One commercially available set comprises a condenser microphone with preamplifier and main amplifier together with a set of ⅓ octave filters and means for switching them sequentially either by manual control or on receipt of an electrical impulse. The ⅓ octave filters may be selected singly or in sets of three to obtain an analysis in octave bands. The outputs from the filters are fed to a chart level recorder incorporating a logarithmic amplifier, so displaying the sound-pressure level in decibels above any chosen reference.

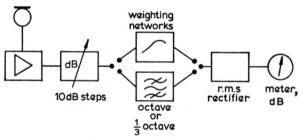

Fig. 2.2 Schematic of sound-level meter

2.2.3 Sound-level meters

For general investigation of noise problems, the most convenient equipment, and the one in commonest use, is the sound-level meter. This is a portable apparatus that is designed to measure the sound level of a noise (see definition in Section 2.1) and is usually adaptable for use in conjunction with a set of octave-bandpass filters to measure band-pressure levels in addition.

The equipment, shown in block form in Fig. 2.2, consists of an omnidirectional microphone, an amplifier with weighting networks

intended to give the apparatus an earlike response, switched attenuators and an indicating meter that shows the weighted-sound-
pressure level. The rectification used is of the root-mean-square
type, to conform to the standards for noise measurement, and is
normally preceded by an attenuator with calibrated steps of 10 dB.
The characteristics of 'precision' and 'industrial-grade' meters are
specified in BS4197:1967 and BS3489:1962.

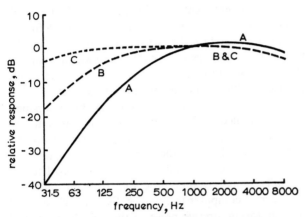

Fig. 2.3   Response/frequency characteristics of A, B and C weighting networks
of sound-level meter (BS4197:1967)

Fig. 2.3 shows the three standard weighting networks used with
sound-level meters. These three are designated networks A, B
and C, and sound levels incorporating these weightings are distinguished from unweighted sound- or band-pressure levels by the
addition of the appropriate letter, e.g. 90 B(a). These networks
were originally designed to simulate earlike response at intensities
of 40, 70 and 100 B, but are now used on an empirical basis
according to circumstances. For instance, the A weighting is found
to give the best agreement with the subjective evaluation of traffic
noise (Robinson, Copeland and Rennie, 1961).

Connection to octave-bandpass filters instead of the built-in
weighting networks is normally effected by means of jacks, the
filters being interposed between the output of the amplifier section
and the input of the meter rectifier circuit.

The chief disadvantage of the sound-level meter as the sole
instrument for use in connection with broadcasting is that there is

no example at present manufactured that is sensitive enough to carry out a band analysis of studio background noise about the maximum permissible levels tolerated for certain programmes. Moreover, for the study of variable noise, such as traffic or aircraft noise, recording for subsequent analysis is the only satisfactory method, and, for this, more versatile equipment, such as that previously described, is more convenient.

2.2.4 Automatic logging of noise measurements

The specialised equipment mentioned in Section 2.2.2 lends itself conveniently to the automatic logging of noise signals, both for the study of fluctuating noise and for the measurement of sound insulation by the method to be described in Chapter 4, in which comparative measurements of sound-pressure levels are made alternately on either side of a partition. The electrical impulses that are used to switch the filters can start a punch driver which controls a paper-tape punch. The amplified microphone signals are converted to logarithmic form by the chart recorder, and the excursions of the pen are converted into voltages, which are fed to a digital voltmeter, the output of which, in serial digit form, is passed to the paper-tape punch. The punched tape is processed to obtain the desired information by normal computer techniques. Systems of this kind are in use for monitoring aircraft noise in the neighbourhood of airports (Gracie, 1968) and for many other purposes.

In this scheme, the purpose of the level recorder is solely to provide a logarithmic conversion. This may be achieved alternatively by the use of a logarithmic amplifier such as that described by Lansdowne (1964).

2.3 Use of noise-measuring equipment

2.3.1 General noise-background measurements

The permissible levels of background in studios differ according to the purpose for which the studio is being used, but, in every case, the octave-band levels approach within 10–15 dB of the threshold of hearing, particularly at high frequencies. The criteria are given in Chapter 3. The main component of the noise is usually caused by air turbulence in the studio-ventilation system, and this is generally loudest near the inlet or extract grilles. Random noise originating in the other parts of the building usually creates a sound only slightly lower in level than the ventilation noise, and therefore an attempt should be made, in any studio noise investigation, to identify and measure the separate components of the noise.

The apparatus and observers should be in an adjacent room to avoid affecting the noise background in the studio. It is often possible to identify the main sources of noise fairly quickly by first switching off all machinery in the building that is likely to contribute to the studio noise and then to make a series of band-level analyses as they are switched on in succession. This work, for obvious reasons, usually has to be done at night, when the studio centre is closed down.

Five microphone positions in the studio are generally sufficient to enable the sound-pressure level in each band to be determined with an uncertainty of less than 1 B. The sources of noise that are directly associated with the studio should be added last. In each measurement, the band-pressure measurements for the five micro-phone positions should be converted into power ratios, averaged and reconverted into decibels. If the total spread of readings is not more than 5 B, the arithmetic average of the levels is sufficiently accurate. Table 2 gives power ratios corresponding to each 1 dB step relative to an arbitrary level.

Table 2.2   Power ratios corresponding to 1 dB intervals

Difference, dB	0	1	2	3	4	5	6	7	8	9	10
Power ratio	1	1·3	1·6	2	2·5	3·2	4	5	6·3	7·9	10

If the addition of a source of noise at any stage causes a rise of M decibels in the level, Table 3 shows the actual amount, namely N decibels, by which the new source of noise exceeds the total previously present.

Table 2.3   Level of added component relative to background noise required to give specified increase in total level

Change M in background level, dB	1	2	3	4	5	6	7	8
Level N of new component, dB	−5	−2	0	1·7	3	5	6	7·2

B*

More elaborate means have been used for identifying sources contributing to the background. Information on the direction can be obtained by stereophonic listening; this method was used for distinguishing the direction of travel of underground trains during noise measurement in the shell of Broadcasting House extension before the installation of the studios. The 'stereo microphones' were accelerometers spaced apart on the structure of the shell. Goff (1955) also showed how correlation of the studio noise with the output of microphones placed in sequence near each possible source of interference could be used to identify sources that could not be conveniently switched off.

2.3.2   Noise surveys in unfinished buildings

If there is a noise hazard against which it is necessary to design the sound insulation of a proposed studio, it is often not enough to make measurements of airborne noise in the site, as will be seen from the case of the Broadcasting House extension mentioned above. The source of noise was an underground railway passing close to one end of the basement floor in which ten studios and their associated control rooms were to be erected. The shell in which the measurements were made was very reverberant and not partitioned in any way. The underground-train noise and vibration were therefore distributed fairly uniformly throughout the space, and bore no relation to the local situation for each studio.

The sound level in the finished studios in such a case will be determined by the levels near the interior of the shell adjacent to the studio site and the protection afforded by the walls, floor and ceiling of the studio itself. Measurements of the near-field sound-pressure level in the shell will be seriously affected by the reverberant sound and radiation from surfaces that will later be screened. The only satisfactory way, therefore, is to make measurements of the velocity amplitudes of the surfaces themselves and to calculate from these the expected near-field sound-pressure levels. It is usual to assume, to be safe, that all the energy imparted to the air in contact with the vibrating surfaces is radiated efficiently, so that we may write

$$W = \rho c V^2$$

where W is the power radiated from unit area, V is the r.m.s. particle velocity, and ρc is the characteristic impedance of a plane wave in air (Section 1.2). This method of estimating the sound field in an incomplete studio has been extensively used by Ward

(1962). A fuller account of this and related techniques will be given in Chapter 6 in connection with structure-borne transmission.

2.3.3 Measurement of noise emitted by studio equipment

The emission of noise by equipment such as film and television cameras, clocks, scenery hoists and revolving stages, must be checked before they go into service. If this can be done in the studio in which they are to be installed, there will be no difficulty, since the methods described above for ventilation systems can be used and the effect of studio reverberation is included.

For mobile equipment that is to be used in any of several studios, however, a manufacturing specification is required in terms of the sound emitted by the machine. Adherence to the specification could be checked by any of the methods described in BS4196:1967; but, for this purpose, either a free-field room or a good reverberation chamber with very low background-noise level is required. Broadcasting-echo rooms are never large enough for accurate diffuse-field testing

The most practical method in the absence of special rooms is to choose a well damped and insulated room, such as a small television studio, which can be made available from time to time for short periods as a test room. The reverberation time of the room is measured over the range of testing frequencies, and, from this, is calculated the room contant R', which is defined by the equation:

$$R' = \frac{S\bar{\alpha}}{1-\bar{\alpha}}$$

where $\bar{\alpha}$ is the mean absorption coefficient of the surfaces (see, for example, Beranek, 1964). The calculation of $\bar{\alpha}$ from the reverberation time and the dimensions of the room is carried out by means of the equation

$$T = \frac{0{\cdot}161\,V}{S\log_e(1-\bar{\alpha})} \quad \text{(SI units)}$$

where V and S are the volume and surface area, respectively, of the room and T the reverberation time. This is derived in Chapter 7.

The proportion of direct sound to total sound is $k/(1+k)$, where k is the ratio of the intensities of the direct and reverberant sounds at a distance r from the source. k is given by $R'/16\pi r^2$ (Chapter 4). If the total sound level from the equipment is measured at the distance r, the level due to the direct sound only may therefore be found by subtracting a correction given by

$$10 \log_{10}(1+1/k)$$

from the measured sound level. In a small studio at the BBC Television Centre, London, which was used for the purpose, the correction was 1 dB at a distance of 1 meter, rising to 3 dB at 1·8 m. It was found practicable to do measurements, including measurements of the directional characteristics of noisy equipment up to a distance of about 1·5m; beyond that distance, corrections for both ambient noise and reverberant sound became too large for an accurate assessment of the equipment.

2.4   Derivation of loudness and loudness level from band-pressure levels

Chapter 1 dealt with the varying sensations of loudness produced in the hearer by tones of the same intensity but of differing frequency, and Fig. 1.6 showed the relationship in graphical form for both pure tones in a free field and for narrow bands of noise in a diffuse field. If two or more tones or discrete bands of noise are presented simultaneously to the ear, the loudness level of the combination cannot be found by simple summation; it requires an intermediate process using an arithmetic function of the intensity, which has already been mentioned—the sone. This is a unit of loudness derived from subjective experiments, in which the subject is required to judge when two tones give loudness sensations in the ratio of 2:1. The larger loudness is then said to have twice the loudness, in sones, of the other. A fixed point of the scale is provided by the definition that a tone of 40 phon loudness level has a loudness of 1 sone. Thus, if S is the loudness, and P is the loudness level in phons, we have

$$S = 2^{(P-40)/10}$$

To perform a loudness summation for a combination of tones or narrow bands of noise, the method is, in principle, to convert the separate phon values into sones, to add them up, and then to convert the sum back into phons. A simple procedure depending on this principle was first introduced by Mintz and Tyzzer (1952) but has been improved on by further research. Two methods have now been thoroughly established, and are recommended as alternative procedures in BS4198:1967.

In the first method, devised by Stevens (1956), the starting-point is an analysis of the noise in octave bands. The contribution of each band is converted into a number, the loudness index, which is determined by the band-pressure level and the geometric mean frequency of the band; this process replaces the successive conversion of each band level into phons and sones. Finally, the total

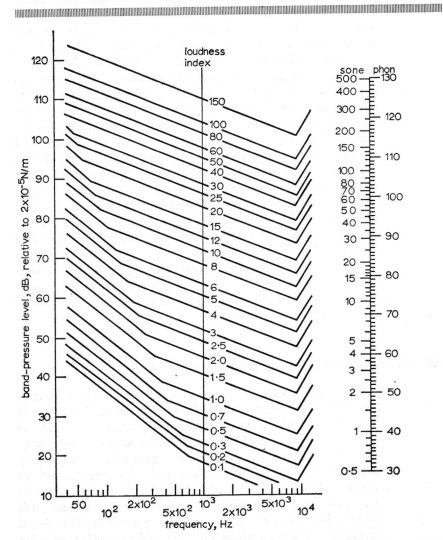

Fig. 2.4 Chart for derivation of loudness indexes, octave bands (Stevens, 1956)

loudness is computed by adding the greatest of the loudness indexes to 0·3 times the sum of the rest of the indexes. The factor of 0·3 represents the diminution, in total loudness, of the several bands resulting from mutual masking; the loudest band is not so multiplied, on the reasonable assumption that the loudness of the noise will not be less than that of its loudest band alone. Fig. 2.4 is the chart published by Stevens for the derivation of loudness indexes.

The alternative recommended method is that of Zwicker (1960). This starts from a ⅓ octave analysis of the noise spectrum, and accounts accurately for the interband masking. As mentioned in Chapter 1, ⅓ octave bands correspond closely to the critical bands over most of the audible frequency range, though, at low frequencies, below about 200 Hz, the critical bands become wider and are represented in the Zwicker procedure by two or three ⅓ octave bands taken together. The separate contributions of the approximate critical bands are converted into sones and drawn as a histogram on specially prepared graph paper. A portion of such a histogram is shown in Fig. 2.5; masking between adjacent bands is taken into consideration by drawing lines from left to right, starting at the right-hand side of each band and parallel to prepared lines, such as AB in the figure. The area included between the curved line so drawn and the histogram represents the extent of the interband masking.

Finally, the mean height in sones of the diagram above the baseline is computed with the help of a planimeter, or simply estimated by eye, and reconverted to loudness level in phons by means of the equation already given.

2.5 Characteristics of noise related to loudness

The loudness of a sound is not *per se* the most important characteristic from the point of view of subjective effect. If the noise is clearly audible above the background, the effect on the listener will generally increase with the loudness, but may be considerably aggravated by features of its spectrum or nature that do not directly contribute to loudness. Thus noise carrying information, such as speech or characteristic types of electrical interference, may be more disturbing on a programme than random noise without such features. In the wider field of community noise, it has been found necessary to compare noises for their subjective effect by using weightings specially devised for the purpose. The suitability of the A weighting for the assessment of traffic noise has already been

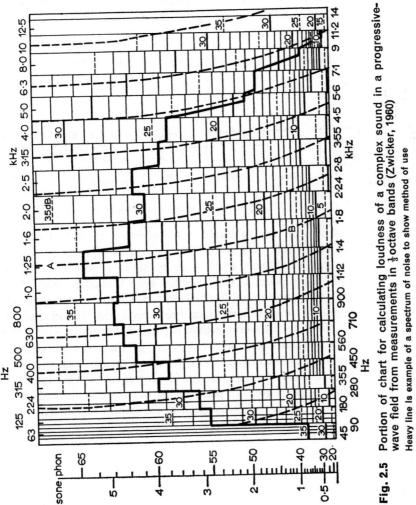

Fig. 2.5 Portion of chart for calculating loudness of a complex sound in a progressive-wave field from measurements in $\frac{1}{3}$ octave bands (Zwicker, 1960)

Heavy line is example of a spectrum of noise to show method of use

mentioned; aircraft noise is assessed by conversion to a single figure either by the use of a special weighting network in a sound-level meter that gives greater weight to both the low and the high ends of the frequency scale [dB(P)] or by a process similar to that of Stevens (1956) similarly modified, giving 'perceived noise dB', or PNdB (Kryter, 1959). A comparison of the various methods of loudness summation was given by Zwicker (1967) using a wide variety of noise samples.

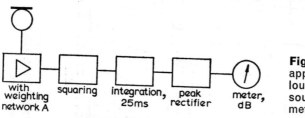

with
weighting
network A   squaring   integration,
25ms   peak
rectifier   meter,
dB

Fig. 2.6 Block diagram of apparatus for measuring loudness of impulsive sounds by Niese's (1965) method

A special case of loudness measurement is that of impulsive sounds, which yield lower relative figures of loudness level by summation of band levels than by direct subjective comparison with steady sounds. This case has been studied by Niese (1965) and others, and sound-level meters adapted to the measurement of impulsive sounds have since been developed. In principle, the output of a square-law rectifier is partially integrated with a 25 ms time constant, and the peak value of the resulting voltage is displayed by the meter. Figure 2.6 shows the block diagram of the Niese method. The time-integration corresponds to the integration which appears to take place in the hearing process and ensures that the correct weight is given to the impulsive sounds.

Chapter 3

Background noise in broadcasting studios

3.1 General considerations

3.1.1 Radiation of noise into a studio

The background noise in a studio is made up partly of energy generated outside the studio and partly of internally radiated energy. Measures for improving quietness must be directed towards both. With internal sources of noise, the only practicable method is to reduce the sound at source; external sources may be treated either by reducing their inherent sound output or by reducing the transmission of sound from the source to the interior of the studio.

In either case, the sound that is actually radiated into the studio is picked up by the microphones together with sound reflected by the surfaces of the room (the reverberant sound).

Let us assume that the source radiates sound power P into the studio. The source in question may be a piece of studio equipment, or it may be an area of wall through which sound is being transmitted from an outside source. Anticipating Chapter 8, we will assume that, on the average, a fraction α of the sound energy falling on any surface in the studio in a period of time is absorbed and that the rest is reflected back into the studio.

Let the mean intensity of the sound in the studio be denoted by I. Of this, a component part I_r is sound that has been reflected at least once, and therefore makes up the reverberant field. The fraction arriving at a point in the room directly from the source has an intensity I_d.

Consider a wavefront forming part of the reverberant sound. This travels across the space at a speed c and is reflected repeatedly from the surfaces at mean intervals t_m given by

$$t_m = l_m/c \qquad \ldots \ldots \quad (3.1)$$

where l_m is the mean free path.

Kosten (1960) has shown that, for an enclosure of any shape (with some restrictions that need not concern us here),

$$l_m = 4V/S \quad \ldots \ldots \quad (3.2)$$

where V is the room volume and S is the total area of its boundaries. The total reverberant energy in the studio is VI_r, and thus the rate of absorption of reverberant energy by the walls, floor and ceiling is

$$VI_r \alpha c S/4V \quad \ldots \ldots \quad (3.3)$$

The 'direct' sound, of power P, suffers a loss $P\alpha$ on its first reflection, leaving $P(1-\alpha)$ as the reverberant-sound power; this is the sound energy that is dissipated in unit time, so that

$$P(1-\alpha) = VI_r \alpha c S/4V . \quad \ldots \ldots \quad (3.4)$$

whence

$$I_r = 4P(1-\alpha)/cS\alpha \quad \ldots \ldots \quad (3.5)$$

Now, assume that the source radiates power all round, so that the power passing through unit area at a distance r from the centre of the source is $P/4\pi r^2$. This assumption is clearly tenable for some sources, but, in others, e.g. an area of wall radiating transmitted sound, it would need to be modified.

The intensity at this distance due to the direct sound is therefore given by

$$I_d = P/4\pi r^2 c \quad \ldots \ldots \quad (3.6)$$

The total intensity at the point is, therefore,

$$I = I_d + I_r = \{P/4\pi r^2 c + 4P(1-\alpha)/cS\alpha\} \quad \ldots \quad (3.7)$$

Now, as shown in Section 1.1,

$$\text{intensity} = p^2/\rho c^2$$

whence the sound pressure at the point is given by

$$p^2 = \rho c P\{\tfrac{1}{4}\pi r^2 + 4(1-\alpha)/S\alpha\} . \quad \ldots \quad (3.8)$$

From Section 1.1,

$$10 \log \{p^2/p^2_{ref}\} = \text{sound-pressure level (s.p.l.) and}$$

$$10 \log_{10}(P/P_{ref}) = \text{power level, } L_P$$

where $P_{ref} = 2 \times 10^{-5} \ N/m^2$, and $P_{ref} = 10^{-12} \ W$.

Hence, finally, by taking the logarithm of both sides of eqn. 3.8,

$$\text{s.p.l.} = L_P + 10 \log_{10}\{\tfrac{1}{4}\pi r^2 + 4(1-\alpha)/S\alpha\} + 0\cdot5 \text{ decibels (SI units)}$$
$$(3.9)$$

The factor $(1-\alpha)/S\alpha$ occurring in the last term will be recognised as the reciprocal of the room constant defined in Section 2.3.4, and denoted as R', following Beranek. The ratio of the direct to the reverberant sound will be seen to be $R'/16\pi r^2$, the result that was also quoted without proof in the same Section.

3.1.2 Noise surveys

The first step in the design of a studio on a new site should be a noise survey of the site to ascertain the noise spectrum and vibration levels in the ground and to identify all important sources of noise. If the studio is to be in an existing building, the vibration levels should be measured on all the walls as well as the floor. Details of vibration measurements and their interpretation are given in Chapter 5.

Noise due to aircraft, road traffic and other external sources must be measured over a considerable period, to take in all normal variations. The object is to design walls and roofs with sufficient sound-transmission loss to reduce the statistical probability of interference to an agreed level. Noise expected from other studios and technical areas in the same building must receive the fullest and most detailed study in the earliest stages of planning (Chapter 10). Finally, the radiation of noise by studio equipment, especially the ventilation system, must be taken into account.

To keep to a minimum the expense of reducing noise at source and of providing sound insulation and vibration isolation, one important task is to decide, as accurately as possible, how great a noise level can be tolerated in the studio without impairment of the programme. This will be considered in the next Section 3.2.

3.2 Methods of deriving criteria of acceptability for background noise

3.2.1 Direct experimental determination of criteria

The audibility of background noise in a broadcast programme depends on the nature of the unwanted noise and on the maximum and minimum intensities of the wanted sound. The intensity of the wanted sound depends in turn on the strength of the source and the distance of the microphone from it, and the nature of the noise might be described in terms of its spectrum, intensity, continuity and information content. The information-content factor depends on whether it is meaningless noise, such as, for instance, ventilation noise, or noise having a recognisable character, such as speech or the sound of a motor horn.

It would seem to be a reasonable assumption, for a start, that the level of background noise that would be just acceptable for a particular programme could be predicted in terms of these variables, and many attempts have been made to do so.

One such investigation, carried out by the author and his colleagues at the BBC research department, may be worthy of mention because, although it did not succeed in its primary aims, it provided related data that later found application. The experiments consisted of listening tests designed to discover the following data: (*a*) the least tolerable ratio of speech to background levels, the noise being confined to octave bands; (*b*) the way in which noise in the several octave bands were combined; and (*c*) the effect of the type of programme and the information content of the noise.

Subjects were asked to classify noise superimposed on recorded extracts of speech or music into five categories, ranging from 'imperceptible' to 'very disturbing'. This was carried out, first, by successive presentation of samples of combined noise and programme for judgment, and, secondly, by asking the subject to combine the noise recording and the programme by means of a 2-channel mixer in such proportions as to produce various degrees of subjective impairment. It should be said at once that the subject used uncalibrated controls in this operation, the chosen settings of the attenuators on the mixer being remotely monitored by the experimenter.

Both methods gave substantially the same results. The conclusions were as follows:

(*a*) The audibility of the noise depends not only on the signal/noise ratio but also on the absolute noise level. To express this in figures, if the level of the programme is increased by 6dB, and that of the background noise by 3dB, approximately the same degree of disturbance will be given. This explains the common observation that the background noise on a programme may be heard and judged more easily if the monitoring level is raised, although the signal/noise ratio is unchanged

(*b*) For a given level of the programme, the curve of the greatest signal/noise ratio against frequency follows the curve of ear sensitivity near to the threshold of hearing

(*c*) Absolute signal/noise ratios could not be assigned to particular studios or programmes because of the number of variables unsuitable for numerical expression.

Further experiments failed to show any clear influence of information content in the background noise. The reason was undoubtedly that the various types of noise were accompanied by differences in spectrum and crest factor, which led to differences in the objective values assigned to them being greater than the subjective differences.

3.2.2 Determination of background-noise criteria from operational experience

The difficulty of analytical study of background noise has led to the establishment of criteria largely by a long process of comparison of measured sound levels with agreed subjective assessments of particular broadcasts. This has been done with fairly concordant results by Beranek (1957), Kosten and van Os (1962) and the present author (Gilford, 1967). In the first two instances, the recommendations take the form of curves representing a series of spectra in octave bandwidths from which one appropriate to a particular requirement can be selected. The present author's curves were restricted to three that were found suitable for studios classified into three corresponding groups. These are

> group 1: sound studios for light entertainment
> group 2: all television studios;
>   all sound studios except in those groups 1 and 3
> group 3: sound-drama studios.

These curves (Fig. 3.1) were arrived at simply from an examination of the records of background noise accumulated over many years of regular measurement and classification of the studios concerned into 'satisfactory' or 'excessively noisy' for the various types of programme. In due course, a consistent pattern emerged that enabled curves to be plotted relating permissible s.p.l. to frequency, normally in octave bands, for the three classes of studio. Most of the older records that were analysed for the purpose were measured with the aid of a fast-rise quasi-peak-reading meter, and were converted into approximate r.m.s. values by the use of the corrections based on measurements of random noise, as shown in Table 3.2. It is likely, therefore, that impulsive sounds of low pulse-repetition frequency will have been slightly overweighted in the results as a whole, leading to conservative criteria.

The derivation of criteria of acceptability by compiling the results of experience gives little or no information on the relative importance of components of different frequency. This is because curves of background noise comprising the objective information must be assessed subjectively as a whole, and the types of noise making up the background tend to have similar spectral shapes, which are reflected in the criterion curves. For this reason, the BBC criterion curves, shown in Fig. 3.1, are of similar shape to those of Beranek and of Kosten and van Os previously referred to.

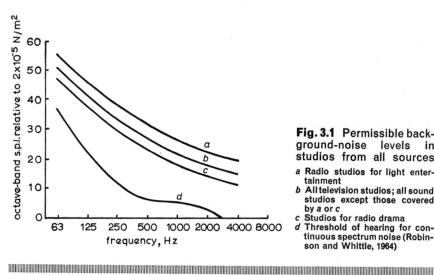

Fig. 3.1 Permissible background-noise levels in studios from all sources

a Radio studios for light entertainment
b All television studios; all sound studios except those covered by a or c
c Studios for radio drama
d Threshold of hearing for continuous spectrum noise (Robinson and Whittle, 1964)

Criterion curves of considerably lower level throughout the frequency range have been published by Kuhl (1964), and are used by broadcasting organisations in West Germany. The OIRT (Organisation Internationale de Radiodiffusion et Television) uses similar curves in eastern-European centres. They follow the threshold of hearing from low frequencies up to the point at which the inherent noise from the quietest of existing microphones becomes the dominant source of programme impairment. Satisfaction of these criteria may be said to ensure a complete absence of unwanted noise, but, in the author's opinion, they are unnecessarily stringent. It must be conceded that, for rather rare events, such as performances on the clavichord for which larger-than-life amplification may be used, these very stringent criteria are justified. However, it is not economical to apply such expensive specifications to all studios when only one or two, suitably treated, would be sufficient to cover all bookings of such a sensitive nature, and therefore uneconomical, since extreme measures for silencing equipment or excluding external noise from a studio are expensive.

3.2.3   Effects of programme-meter characteristics and other factors

An interesting series of experiments using artificially generated noise in programme circuits has been described by Geddes (1968). The background noise ranged from random noise and regular impulses to carrier crosstalk and hum. Listening tests were made both in monophonic and in stereophonic sound. Each sound was subjected to four different frequency weightings in turn, including the A weighting, and the levels at which they became subjectively equal as impairments were measured with seven different programme meters using different methods of rectification. The most consistent of the ratings was given by a meter currently proposed by the OIRT and making use of the integration circuit (Fig. 2.6) introduced by Niese (1965), together with the same square-law integration. Unlike Niese's sound-level meter, the integration was preceded by a modified weighting characteristic proposed by the Deutsches Bundespost for the CCITT (Consultative Committee for International Telephones and Telegraphs). This circuit gave the most constant reading for the series of noises giving the same subjective impairment to speech or music. Geddes found the impairment threshold for a wide range of noises to correspond to a measured signal/noise ratio of 60 dB, the figure being the same within 2 dB for stereophonic as for monophonic listening. He did not point out any systematic difference between the impairment thresholds of noises with and without information content, though an examination of his results suggests that substantially random and unpitched noises may be about 3 dB higher in level than those having a content of speech or tone for equal subjective rating.

He also examined the typical gain settings, used in microphone amplification circuits for music and speech, and concluded that acoustic noise originating in the studio was more likely to be intrusive with live speech than with music if the curves of Fig. 3.1 were adhered to. This does not appear to be borne out in practice, but, if correct, it would suggest that speech studios for sound should be included in group 3 and that sound drama studios should follow a curve approximately 4 dB lower than the group 3 curve.

It will be seen that there are some contradictions between the recommendations of the various workers mentioned above, and Fig. 3.2 shows the extent of these deviations. The four curves represent the criteria for a sound-drama studio as derived from the recommendations of Kosten and van Os, Kuhl, Gilford and Geddes.

The differences between the criterion curves in Fig. 3.2 are, no doubt, partly consequences of differing techniques of production between the organisations using them; e.g. the use of more directional microphones and closer working positions will allow higher background levels to be tolerated, whereas the use of compression in the programme chain will necessitate a lowered background noise, since the gain of the chain will increase during gaps in the programme.

In what follows, references to criterion curves will be taken to indicate those of Fig. 3.1, and the data in Chapter 4 on sound-insulation requirements are all based on these curves, which have proved generally adequate.

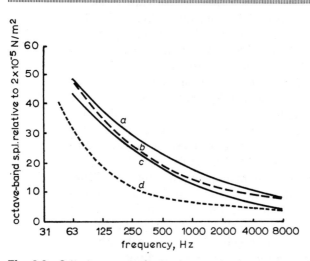

Fig. 3.2 Criterion curves for background noise applicable to radio-drama studios
a Gilford (1967)
b Kosten and van Os (1962)
c Geddes (1968)
d Kuhl (1964)
s.p.l. = sound-pressure level

3.2.4 Control and listening rooms

It is found by experience that the background-noise level in a room for listening, mixing or control should not exceed that specified for the studios with which it is linked.

3.2.5 Studios for rehearsal or temporary use only

It is often suggested that money could be saved, in the construction of studios for rehearsal only or for temporary use, by relaxing specifications for background noise. It is argued that, for a studio required for rehearsal only, background noise may be largely disregarded because it will not be heard on the transmitted programme. For studios for temporary occupation, it is similarly thought that a noisy ventilation system can be tolerated for the time being, as it can be used for rehearsals only and switched off immediately before transmission or recording. This is a very unsatisfactory policy. Excessive background noise during rehearsals, particularly of television productions, causes avoidable equipment sounds and other production faults to escape notice until they appear on transmission. Nor is it satisfactory to operate noisy ventilation plant to 'blow out' a studio between periods of rehearsal and recording, because, sooner or later, a rehearsal will run too long and have to be abandoned to cool the studio, or recording schedules will be upset for the same reason. Thus a temporary studio should be brought as near as possible to the same standards of quietness as are necessary for a permanent facility. As a general rule, noise curves not more than 6dB above those of Fig. 3.1 should be the aim in equipping a studio that is for brief temporary use or a studio that is not intended for use for transmission or recording.

3.3 Specifications for individual sources of noise making up a total background

The background noise in a studio will, in general, be composed of continuous sounds that are always present while the studio is in use, and sounds, such as those of certain types of studio equipment, that occur only now and again.

If all these components were regarded as independent and equal in importance, it would be necessary to maintain each component well below the permitted total represented by the curves of Fig. 3.1, so that, if they all occurred simultaneously, they would not intrude on any programme. However, provided that the total continuous noise and each of the intermittent components are each 3dB or more below the permitted total, the resultant background is generally satisfactory, because the intermittent components coincide only infrequently and, in any case, may have characteristics very different from one another.

Among noises of the continuous type, which are present as a constant feature when the studio is in use, are those of ventilation machinery and air turbulence, hums from power supplies, camera-cooling blowers, line-scan whistles and certain kinds of back-projection equipment. All these sounds should, as explained above, be considered as part of the general background, and they should together lie at least 3 dB below the curves of Fig. 3.1, so that an added intermittent noise of equal individual intensity will not raise the total above the criterion.

Intermittent noises are those of scenery and lighting hoists, revolving stages, the motors of camera dollies, and trailing camera cables.

The noise of equipment operating in a studio is of particular relevance to television, because a great deal of powerful and mobile machinery is an indispensable adjunct of television productions. In contrast, radio studios contain little equipment that is likely to cause disturbance. In television studios, every item of equipment must be the subject of manufacturing and acceptance specifications to ensure that it shall produce no more than the permitted noise level at an operational microphone.

3.4 Noise of particular items of studio equipment

3.4.1 Acceptance tests

Methods of making measurements to establish whether a new piece of equipment satisfies its specification for noise output were described briefly in Chapter 2. For such measurements, the nearest operational microphone is assumed to be at a distance of 1·3 m.

3.4.2 Ventilation systems

If the ingress of external noise is assumed to have been satisfactorily reduced by effective sound insulation and by isolation from structure-borne sound, the major source of background noise is usually the ventilation system. A large studio for colour television may be equipped with lights consuming up to about a third of a megawatt of power, and this power is almost entirely converted into heat, which raises the temperature of the air. If the ventilation fails in a studio such as this, work becomes impossible within about 15 min if the normal production lighting is left on. If we assume that a rise in temperature of 15°C is permissible as the ventilating air flows from inlet to exhaust ducts through the studio, the mass of air to be moved per second is given by

$$300\,000/15C_p \text{ kilogrammes}$$

where C_p is the specific heat of air at constant pressure. This represents a volume of $3 \times 10^5/15\rho C_p$ cubic metres, giving approximately 6×10^4 litres or $60\,\text{m}^3/\text{s}$. This is clearly a very large flowrate, demanding large ducts and a high velocity within them.

Fig. 3.3 shows a typical ventilation installation for a large television studio. Air is drawn from the outside of the building through filters by a large centrifugal fan and discharged into a mixer chamber from which ducts lead to the upper part of the studio.

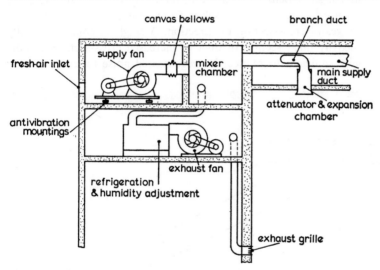

Fig. 3.3 Typical ventilation system for large studio, showing main components

The fan is isolated from the floor by antivibration mountings, or by a simple slab of cork serving the same purpose, and is connected to the mixing chamber by a short section of flexible trunking. A second fan draws air from exhaust ducts in the studio, and discharges it into the mixing chamber by way of a refrigeration plant. Automatic controls maintain the proportion of recirculated air, and thus keep the air supply to the studio at a desired temperature. The air is conveyed from this mixing chamber to distribution ducts in the upper part of the studio. These ducts are branched off a main duct running along one side or across the middle of the studio, thus distributing the air evenly all over the area. Alternatively, the air can be ducted to the perimeter of the ceiling, from where it sinks towards the floor, taking the place of that being drawn out of the studio by the exhaust ducts near ground level.

The inside of the mixer chamber and the larger-section parts of the ducting are lined with a porous, sound-absorbing material, which reduces the transmission of sound from the fans along the ducts to the studio. The arrangement of the inlet ducting is designed, as far as possible, to cause the least turbulence or localised flow in the neighbourhood of microphones; this is most important in sound studios in which distant microphones are commonly used.

To move the air at a sufficient rate, it may be necessary to use fairly high velocities of flow, and there will be some regeneration of noise in the ducts themselves from turbulence, particularly at changes of cross-section or direction. A single attenuator, usually a heavily lined section of duct of increased cross-section, is normally sufficient to reduce this regenerated noise to a satisfactory

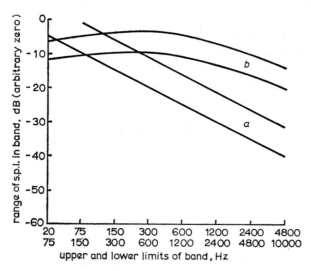

Fig. 3.4 Spectra of noise of ventilating fans (after Beranek, 1960)

a Centrifugal fan
b Axial fan
Upper and lower lines for each fan show approximate limits

level. The attenuator may contain lined splitters (metal sheets dividing the duct into several parallel channels), or it may be in the form of a labyrinth.

Methods of calculating the noise output of a ventilation system have been given by several authors, notably Beranek (1960). The basis of these methods is to compute the total production of sound energy in the fan by means of an empirical formula that includes the nominal power of the driving motor and the static pressure produced or the volume flow of air through the system. This total power is then fitted to one of a series of spectra for the types of fan that are available (Fig. 3.4). Centrifugal fans produce relatively more low-frequency noise than axial blowers. These are less frequently used for large ventilation systems, but are feasible for systems using high velocities and small ducts.

The reduction or regeneration of sound power as the air is conveyed through the ducting to the studio is then examined in detail, taking into account the effect of bends, duct linings, changes of section and the use of attenuators. Finally, the sound-pressure level at the nearest point of the studio to each outlet duct at which an operational microphone is likely to be situated is calculated, using an equation such as eqn. 3.9.

A point of importance in the design of such a system is the effect of the grilles that are often fixed over the ends of ducts where they terminate at the studio walls. These are used to regulate, or modify the direction of, the airflow, to prevent papers or cigarette ends from being drawn into exhaust ducts and for purely decorative purposes. Unfortunately, if not well designed, these can be the source of considerable regeneration of noise, either by turbulence or by the production of clearly pitched aeolian tones originating from a regular series of vortexes that are shed by the vanes or bars of the grilles and travelling down the airstream. Generally speaking, such grilles serve very little purpose in studios, and are best avoided if not fulfilling a definite function.

3.4.3 Noise of electrical-power and other equipment

A considerable amount of low-frequency hum, most commonly at the harmonics of the 50 Hz supply frequency, can be generated by electrical equipment associated with lighting and other services. To maintain constancy of lighting output and long life of the lamps, regulated voltage supplies are necessary, often incorporating saturated transformers. These automatic voltage regulators may be sources of low-frequency vibration, and must therefore be located and mounted so as to avoid transmission of vibration to the walls of the studio. At the same time, it will be an advantage to mount them as near as possible to the studio to reduce the lengths of heavy high-power cables.

It is normally found sufficient to mount the regulator on rubber antivibration mountings giving a transmission loss of about 20 dB at the relevant frequencies; heavy armoured cables carrying the unregulated supply and regulated output should not be allowed to provide direct links between the regulator and the structure of the studio (Fig. 3.5).

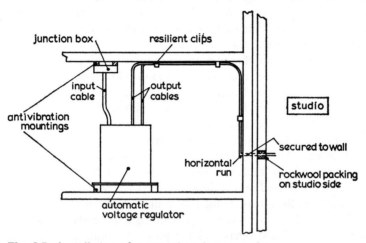

Fig. 3.5 Installation of automatic voltage regulator

Another cause of hum at harmonics of the supply frequency is the use of thyristor-controlled lighting. In this method of control, the lighting current is passed through thyristors controlled by biasing voltages, which serve to switch the current at a point on the supply waveform that may be altered to vary the mean current over the whole cycle (Fig. 3.6). The risetime of the current following each switch on is of the order of $2\mu s$, and such a rapid rise may shock-excite mechanical components in the lamps, causing hums with very high harmonic content. Moreover, the rapid flux changes, in association with any wiring loops in the studio, will induce signal voltages in electromagnetic microphones that have not been effectively degaussed. Both these undesirable effects can be reduced by 20 dB or so by the insertion of carefully designed filters into the lamp circuits.

3.4.4 Noise from scenery and lighting hoists

One feature distinguishing television studios from radio studios is the occasional necessity to move heavy objects, such as scenery components and banks of lights, about the studio during transmission or recording. Such operations are carried out by electric winches installed on the lighting grids over the studio or on the upper parts of the walls. Noise from the motors may be reduced to a satisfactory level for operation during productions by enclosing each motor in a heavy wooden box lined with efficient sound-absorbing material. Radiation of low-frequency sound from the box will be inefficient, but it must be isolated by means of anti-vibration mountings from any surface, such as the studio wall, to which it is fixed and which could otherwise improve the impedance match between the box and the air.

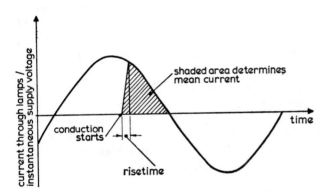

Fig. 3.6 Illustrating control of lighting by means of thyristors

3.4.5 Camera and monitor line-scan whistles

The whistles emitted by the line-scan transformers of television cameras and monitors deserve a special mention, because they can be a source of annoyance and fatigue to the studio personnel. Indeed, some operators have feared damage to their hearing when the intensity was sufficient to cause discomfort.

This very important possibility has been investigated by the BBC research department. The sound-pressure levels of the line-scan whistles in the immediate neighbourhood of a number of representative television monitors were measured at the nearest feasible operating distance. The maximum level near any monitor was 54 dB, whereas, in an area containing several monitors, levels were found as high as 68 dB. As no data existed on the damaging effects of such high-frequency sound (10·025 and 15·265 kHz for 405- and 625-line pictures respectively), volunteers in the department were exposed to tone of these frequencies in a free-field room for 30 min. at a time at levels up to 90 dB. No temporary threshold shifts were found to have taken place in any of the subjects as a result of this exposure, though some of them gained the impression that shifts of the order of 20 dB had occurred. They complained of tinnitus and of hearing random noise immediately after the end of the exposures, yet retained normal thresholds in spite of these sensations.

There is no doubt however that, although no danger to hearing could be established, operators do feel distress on long exposure to excessively noisy sets. Few manufacturers of equipment pay any real attention to reduction of whistle, though very substantial reductions can be achieved by simple precautions in design and construction, such as encapsulating the line-scan transformers. An s.p.l. not exceeding 40 dB at 1·3 m should be specified; this would be inaudible in reproduction on domestic receivers compared with receiver-generated whistles, and would make the lot of the studio staffs more pleasant.

Fortunately, with the gradual obsolescence of the 405-line system, together with the growing use of 625-line cameras with standards convertors as the source of the remaining 405-line transmissions, line-scan whistles will soon be a thing of the past except for the minority who have ears with a high enough cutoff to be disturbed by tone of 15 kHz frequency.

3.4.6 Holding specifications

It is often not practicable to quote specifications for a room background or for equipment noise that will ensure fully satisfactory broadcasts or listening conditions. Examples are found in mobile control vehicles or studios, for which it is impossible to provide adequate sound insulation because of weight limitations, and in studio equipment that is necessary for programme productions in spite of being inherently noisy. In the absence of any specification, there would be no control on manufacture or installation, and it is therefore essential to establish a 'holding specification' that represents the best of current practice, together with specified improvements for subsequent generations of the same equipment. These specifications must be drawn up for each case, taking into account all possibilities of modification and redesigning for future production.

Chapter 4

Airborne-sound insulation

4.1 Introduction

4.1.1 Introductory concepts

The background-noise levels appropriate to broadcasting studios and other technical areas are very low compared with external-noise levels due to traffic and other everyday causes and those due to programmes in neighbouring studios. Therefore a decision on the amount of reduction in sound level to be provided by the walls must be a matter of very early priority in designing a studio or broadcasting centre.

The first duty of any author writing on sound insulation is to exorcise that diabolical word 'soundproofing'. It is often used by architects and others to indicate the internal application of materials, usually porous layers or proprietary acoustic tiles, with the object of removing intruding noise, reverberation and audible interference generally. It cannot be overstressed that the exclusion of external sounds requires methods and materials completely different from those for damping internally generated sound and reverberation, and that the two processes are largely independent of each other. Most readers must already know this, but it requires to be repeated, *ad nauseam,* because mistakes are still made often in the design of buildings owing to ignorance of the simple principles.

It is true that the absorptive treatment of the walls of a studio may affect the level of noise due to external sources by reducing the reverberant energy, but it also reduces the reverberant energy generated by the programme sources to exactly the same extent. Moreover, it may necessitate a more distant microphone position for programmes using a natural balance, i.e. one in which all, or most of, the reverberation comes from the natural acoustics of the studio and which aims to give the impression of normal

C

perspective. Any increase of microphone distance will reduce the level of the wanted sound in relation to the disturbing noise, thus defeating the object of the use of the absorbing material.

Furthermore, as a result of the confusion between the functions of sound insulation and absorption, acoustic tiles or similar porous layers are often proposed as sound-insulating partitions, but a numerical example will make it clear how ineffective they would be. A typical sound-absorbing tile, say, 20 mm thick, has an absorption coefficient of about 0·8 at 2 kHz, which means that about 80% of the sound energy falling on it is lost during the passage through the tile to the wall behind and back again to the surface. This represents a loss of about 7 dB in traversing a thickness of 40 mm, which is about equal to the reduction given by a piece of polythene sheeting only $\frac{1}{2}$ mm thick. Good sound insulation requires a heavy nonporous barrier that will neither pass air nor move under the action of sound pressure; absorption requires a material that will admit air (i.e. a porous material) or move under the sound pressures and absorb energy by internal losses.

4.1.2 Definitions

Airborne-sound transmission is the transfer of sound from one point to another, the sound being generated in the air at the first point and received in the air at the second. In the transmission between two rooms, the first is known as the **source room** and the second as the **receiving room.** The shortest path of transmission, usually through one or more intervening partitions, is called the **direct path**; sound that travels by indirect paths, possibly including substantial distances through solid structure, is said to travel by **flanking transmission.** Sound transmitted through the solid material of the building, particularly from a vibrating source in contact with it, is known as **structure-borne sound.** Fig. 4.1 shows these forms of transmission. Structure-borne transmission will be considered in Chapter 5.

The sound-insulating characteristic of a partition is its **transmission coefficient.** This is the ratio of the sound power radiated into the receiving room to that falling on the source-room side. It is usually denoted by τ. The difference in sound-power levels is known as the **sound-reduction index (s.r.i.)**; this is the figure usually quoted. The two quantities are related by the equation

$$\text{s.r.i.} = 10 \log_{10} 1/\tau \qquad . \quad . \quad . \quad (4.1)$$

In this monograph, the term **transmission loss** will be used in places as a less cumbersome alternative to sound-reduction index.

The reduction of s.p.l. between specified points in the two rooms is known as the **sound-level reduction.**

The **mean sound-reduction index** (or mean sound-level reduction) is the arithmetic mean of the values of the quantity taken at $\frac{1}{3}$ octave intervals in the range 100–3200 Hz.

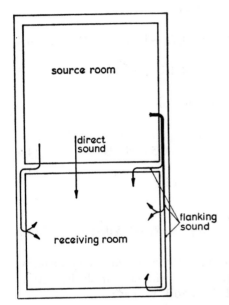

Fig. 4.1 Sound-transmission paths between two adjacent rooms

4.1.3 Relationship between sound-level reduction and sound-reduction index

If the receiving room has reflecting surfaces, the sound-level reduction given by a particular partition will be less than if the surfaces are highly absorbing. This follows from the derivation of eqn. 3.9.

For a flat wall of area A forming all, or most of, one side of the room, the direct component of the transmitted sound consists of a substantially plane wave with an intensity given by P_2/Ac, where P_2 is the transmitted power. Substituting this value into eqns. 3.7–3.9 in place of the term $P/4\pi r^2 c$, we have

$$L_2 = L_P(2) + 10 \log_{10}\{1/A + 4(1-\alpha)/S\alpha\} + 0.5 \text{ decibels} \quad (4.2)$$

If P is the total power output of the source, the reverberant energy falling on the partition is given by AP/R_1 (where R_1 is the room constant of the source room). The s.p.l. L_1 due to the reverberant-sound field in the source room is thus seen from eqn. 3.9 to be

$$L_1 = L_P(1) + 10 \log_{10} 4/R_1 \quad . \quad . \quad . \quad . \quad (4.3)$$

where L_P is the power level of the source.

Combining these equations gives

$$L_1 = L_P(1) + 10 \log_{10} A/R_1 - 10 \log_{10} 4/R_1 \quad . \quad (4.4)$$

$$= L_P(1) + 10 \log_{10} A/4 \quad . \quad . \quad . \quad . \quad . \quad (4.5)$$

Hence the sound-level difference is

$$L_1 - L_2 = L_P(1) - L_P(2) - 10 \log_{10}\{1/A + 4(1-\alpha)/S\alpha\} \\ + 0.5 \text{ decibels} \quad . \quad . \quad . \quad (4.6)$$

But the difference $L_P(1) - L_P(2) = \text{s.r.i.}$

Hence the equation becomes

$$\text{s.r.i.} = L_1 - L_2 + 10 \log_{10}(\tfrac{1}{4} + A/R_1) - 0.5 \text{ decibels} \quad (4.7)$$

If the average absorption in the receiving room is very small, we may regard R_1 as being simply $S\alpha$, without introducing serious errors. If we also disregard the 0·5dB, we have

$$\text{s.r.i.} = L_1 - L_2 + 10 \log_{10}(\tfrac{1}{4} + A/S\alpha) \quad . \quad . \quad (4.8)$$

For highly absorbing receiving rooms, or for radiation into the open air, A/R_1 is small, and we have

$$\text{s.r.i.} = L_1 - L_2 - 6 \text{ decibels}$$

The British Standard for sound-transmission measurements in buildings (BS2750) disregards the $\tfrac{1}{4}$ in comparison with $A/S\alpha$, and recommends

$$\text{s.r.i.} = L_1 - L_2 + 10 \log_{10}(A/S\alpha) \quad . \quad . \quad . \quad (4.9)$$

For the sake of completeness, it should be mentioned that the Standard recommends pragmatic forms of normalisation for comparing walls in dwellings, where there may be multiple paths of transmission; the receiving rooms are regarded as of similar size but different as regards absorption and reverberation time T. The equations are

Normalised sound-level reduction $= L_1 - L_2 - 10 \log_{10} (10/S\alpha)$

and

$$= L_1 - L_2 - 10 \log_{10} (T/0 \cdot 5)$$

4.2 Sound-insulation requirements in studio centres

The permissible levels of background noise in studios for sound and television broadcasting have been specified in Chapter 3, the main recommendations being embodied in Fig. 3.1. A partition between two areas must satisfy the following requirements:

(a) It must reduce the level of interfering sound in each area, so that, when it is combined with unwanted sound generated internally, it does not raise the level higher than the accepted limit.

Generally, transmission in one direction is of more consequence than in the other, and all designs must cater for the direction of the worse potential interference.

(b) It should not be excessively heavy or complex for the purpose, since large factors of safety and inefficient designs are expensive.

(c) If the partition separates areas that are electroacoustically coupled, such as a studio from its control cubicle, the sound reduction must be great enough to prevent howl-round at all frequencies.

The sound-level difference to be created by a particular wall will depend on the level of interfering sound permitted in the 'receiving' area and the sound-pressure level on the other. The sound-reduction index required to achieve that sound-level difference will depend also on the area of the partition and on the absorption in the receiving area, according to eqn. 4.7.

It is necessary to take into account two aspects of certain areas, as sensitive areas and also as sources of interference to others. For instance, a symphony-orchestra studio produces a level of potentially interfering noise comparable with road traffic, so that neighbouring areas must be protected by walls giving very high sound-level reduction.

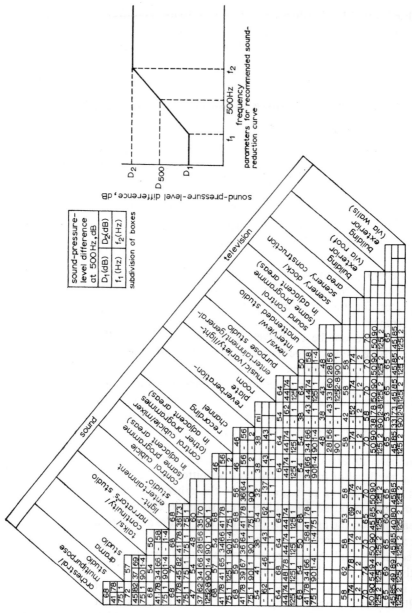

Fig. 4.2 Chart showing sound-reduction parameters for partitions between areas in studio centres (Smith and Gilford, 1968)

* Studios intended for pop-music groups with amplified equipment require partitions giving 10dB higher sound reduction for protection of neighbouring areas.

One must therefore base calculations for the sound-reduction indexes of walls between studios on their expected power outputs as well as on their permissible background levels.

All types of technical, administrative and external areas have been considered in these two aspects in a BBC Research Department Report (Smith and Gilford, 1968), and the results are shown in Fig. 4.2, which is an abbreviated form of the design chart included in that report. The chart is divided into boxes, each of which represents the needed sound reduction between the two areas, and is subdivided into compartments as explained in insets.

It was assumed by the authors that, for noise transmitted from outside, the source was one complete wall of the studio uniformly excited by the transmitted sound, so that eqn. 4.8 could be applied to the calculation of sound-pressure levels. The sound-pressure levels at the source side of the wall were derived, for purposes of calculation, from experimental data on the sound fields associated with sources such as road traffic and aircraft, or from measurements in working areas such as television scenery workshops.

For programme sources such as voices or orchestral instruments, eqn. 3.9 was used, in conjunction with measured or published data on their power outputs, to calculate the expected sound-pressure levels at the walls of the source studio. Where appropriate, the minimum practicable distance of the performer from the nearest wall was taken as r in eqn. 3.9. The room constant R was derived from the mean dimensions and reverberation times of the relevant type of studio.

The correction term $10 \log_{10}(A/R)$ for the reverberant term in the receiving studio varied from -2 to $-6\,dB$ in music studios and from -5 to $-8\,dB$ in television studios.

The level difference to be provided by the wall was calculated at $50\,Hz$ and at every octave from $62\,Hz$ to $8\,kHz$ by simply subtracting the permitted background levels in octave bands from the corresponding levels at the other side of the wall.

4.3 Theory of sound transmission through partitions

4.3.1 Simple walls

It has already been stated that a porous material makes a poor sound-insulating wall. Fig. 4.3 represents a partition of a thickness that is small enough compared with the wavelength of the incident sound, shown to the left, to be regarded as negligible. The wall, being massive and porous, presents a complex mechanical impedance Z per unit area to the incident sound. Let p_i, p_r and p_t be the r.m.s. pressures in the incident wave, the reflected wave and the transmitted wave, respectively, the direction of propagation being, for simplicity, perpendicular to the wall.

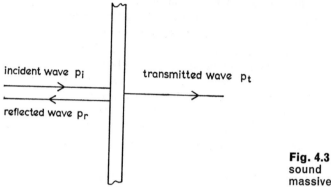

incident wave p_i transmitted wave p_t

reflected wave p_r

Fig. 4.3 Transmission of sound through simple massive partition

Then the volume displacement of unit area of the wall and of the air at its surface is given by

$$v = (p_i - p_r)/\rho c$$

since it is the difference of the particle velocities in the incident and reflected waves. But if we turn our attention to the transmitted wave, we find that this has a pressure

$$P_t = \rho c v$$

Hence

$$P_i - P_r = P_t$$

The displacement of the air-wall combination is also given by the ratio of the pressure across it to its impedance. Hence

$$(p_i + p_r) - p_t = Zv$$

Eliminating p_r and v from these equations, we have

$$p_i/p_t = 1 + Z/2\rho c$$

The square of the reciprocal of p_i/p_t is the ratio of the transmitted and the incident powers, and therefore the transmission constant τ is given by

$$\tau = (p_t/p_i)^2 = 1/(1+Z/2\rho c)^2 \quad . \quad . \quad . \quad (4.10)$$

The impedance Z has parallel components due to the mass of the wall and its flow resistance. Let R_f be the flow resistance and m the mass per unit area, so that the impedance given by the two in parallel is

$$\omega m R_f/(\omega m + R_f)$$

To achieve a high sound-reduction index, m and R must both be high; i.e. the wall must be as heavy as possible and free from porosity or leakages. For a nonporous wall, where R_f is infinite, the equation becomes

$$\tau = 1/(1+m\omega/2\rho c)^2$$

Finally, the sound reduction index is given by

$$\text{s.r.i.} = 20 \log_{10}(1+\omega m/2\rho c) \quad . \quad . \quad . \quad (4.11)$$

which is illustrated in Fig. 4.4. The characteristic impedance ρc

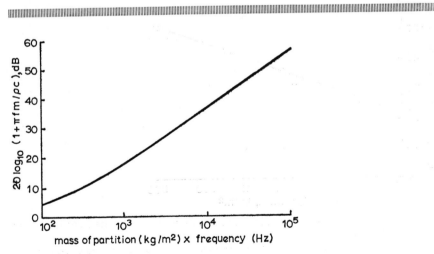

Fig. 4.4 Function $20 \log_{10}(1+\pi fm/\varrho c)$

has a value of 400 SI units at normal studio temperature, and thus for values of ωm over, say, 5000/kg S^{-1}, the first term in the bracket may be neglected. This simplification is valid for any practical insulating partition, and we conclude that the s.r.i. of any limp, impervious partition increases by 6 dB for every doubling of the frequency or the mass per unit area.

4.3.2 Effect of stiffness in simple partition

In practice, the simple mass/frequency law evaluated above is not exactly obeyed, because of the effects of stiffness. Assume that an elastic partition is supported at its edges. Its mass and stiffness will form a resonant system, the lowest mode of which corresponds to the inphase motion of the whole plane relative to its edges. Below this frequency, the motion will be almost entirely determined by the stiffness and the driving force of the sound wave; there will be a minimum of transmission loss at the resonance frequency, above which the transmission loss will rise towards its 'mass-law' value, i.e. the value predicted by eqn. 4.11. As other, higher modes of vibration appear, there will be further minima in the transmission-loss curve. If the incident sound were entirely normal to the plane of the partition, these resonances would appear clearly

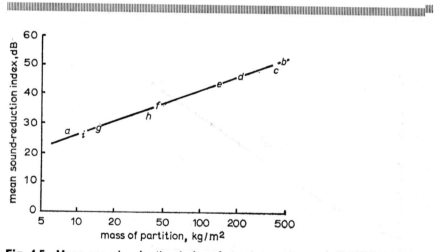

Fig. 4.5 Mean-sound-reduction index of simple partitions plotted against mass unit area

a 3mm glass	d 115mm brickwork	g 25mm wood blockboard
b 180mm concrete	e 80mm clinker block	h 3mm lead
c 230mm brickwork	f 50mm woodwool slabs	i 6mm asbestos-cement sheet

on the transmission-loss/frequency curve, but in a randomly incident-sound field, the fluctuations are largely smoothed out, and the actual curve becomes fairly straight with an increase of about 5 dB for each doubling of the frequency in place of 6 dB for a pure mass. As the stiffness of a partition increases with its thickness and therefore with its mass, the same modification appears in the curve of loss against mass at a fixed frequency. Fig. 4.5 shows the mean values of transmission loss of partitions, averaged at $\frac{1}{3}$ octave intervals of frequency in the range of 100–3200 Hz, plotted against their mass per unit area.

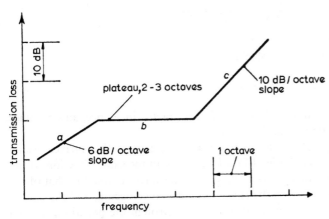

Fig. 4.6 General form of frequency/transmission-loss characteristic of simple partitions (Watters, 1959)

Watters (1959) has shown that the relationship between the transmission loss of a simple partition and the frequency can be represented by a curve consisting of three straight sections, as shown in Fig. 4.6. The section *a* follows the classical mass-law with a slope of 6 dB/octave, *b* is a plateau between 2 and 3 octaves in width succeeded by a section *c* sloping at 10 dB/octave. The height of the plateau *b* is about 26–30 dB for all normal masonry walls.

A more important consequence of the stiffness of a partition was first pointed out by Cremer (1942). When a sound wave strikes a partition obliquely, at an angle θ to the normal, the point of intersection of the wavefront with the partition travels over the surface at a speed given by $c/\sin \theta$, where c is the velocity of sound in air.

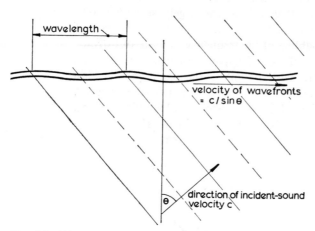

Fig. 4.7 Diagram illustrating coincidence effect

This speed has a minimum of *c* for grazing incidence, and becomes infinite for normal incidence.

The parallel wavefronts travelling across the surface excite bending waves (Fig. 4.7), which travel in the same direction. If the speed of propagation of these bending waves is equal to that of the wavefronts across the surface, a resonance occurs, and the energy of the incident sound travels easily through the partition, causing a dip in the loss/frequency curve. Fig. 4.8 shows the polar diagram

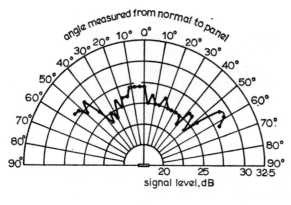

Fig. 4.8 Polar diagram of radiation of 4kHz sound from 4·7mm thick aluminium plate (Burd, 1958)

of the radiation of 4 kHz sound (Burd, 1958) from a 4·7 mm thick aluminium plate. Maxima of radiation, owing to the coincidence effect, will be seen at angles of $+55°$ from the normal.

For a partition between two rooms, in which sound is incident in all direction on one side of the intervening wall, a different type of curve is found. The velocity of bending waves in a strip of material is shown by Cremer (1953) to be

$$c_b = \{EI^2/m'(1-\sigma)\}^{\frac{1}{4}} \quad . \quad . \quad . \quad . \quad (4.12)$$

where
  m' = mass per unit area of the material
  E = Young's modulus
  I = second moment of area of the section about the midplane
  σ = Poisson's ratio for the material.

It is clear that this velocity varies as the square root of the frequency, and is thus highest at high frequencies. At the same time, the velocity of the forcing wavefronts varies with the angle of incidence. Therefore, above the frequency at which the bending-wave velocity reaches the speed of sound in air (the velocity of the forcing waves at grazing incidence), there is a coincidence frequency for every angle of incidence. The frequency at which the speeds are equal is known as the critical frequency, and, above this frequency, the transmission loss is everywhere lower than

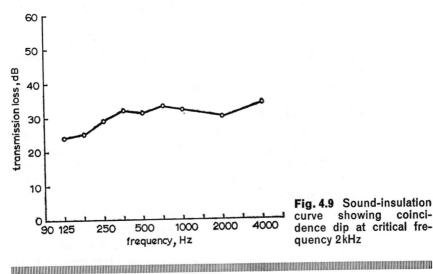

Fig. 4.9 Sound-insulation curve showing coincidence dip at critical frequency 2 kHz

it would be in the absence of coincidence effects. Fig. 4.9 is a typical curve of transmission loss showing this effect.

The critical frequency is found by equating the bending-wave velocity to the velocity of sound in air. From this, we have

$$f_c = (c^2/2\pi)/\{m'(1-\sigma^2)/EI\} \qquad . \quad . \quad (4.13)$$

where f_c is the critical frequency.

f_c is clearly highest for limp, heavy partitions and lowest for light, stiff ones. Table 4.1 shows approximate values of f_c for various common partitions.

Table 4.1   Critical frequencies of common partitions

Material	Thickness mm	Critical frequency Hz
Brickwork	250	120
Concrete	180	103
Brickwork	120	240
Glass	2	7 600
Plywood	3	7 000
Plasterboard	9·5	4 200
Steel	1	12 000

Clearly it is best, in designing a partition of limited mass, to choose materials that have critical frequencies high enough to ensure that coincidence transmission occurs only above the important middle-frequency region.

Westphal (1954) has shown that the critical frequency dip is most severe for lightly damped materials and for unsupported sheets of large area. Reflections at boundaries have the same general effect on the dip as damping.

4.3.3   Multiple partitions

The graph of Fig. 4.5 shows that, if we double the mass of a partition, we gain about 5 dB in transmission loss, provided that we are in a region in which the 'mass law' is obeyed; e.g. a brick wall of 120 mm thickness has a mean transmission loss of 45 dB, and one of 240 mm gives 50 dB. On this basis, a wall intended for a transmission loss of 60 dB, as is often required between studios, would require a thickness of 0·84 m and a mass of approximately 1800 kg/m². On the other hand, it might be expected that two separate walls of 120 mm each would provide a transmission loss of

90 dB. This can only happen, however, if the two walls are so far apart that the sound field at the second wall is completely uncorrelated with that radiated from the first, and this can happen only when they are separated by a distance equal to several wavelengths of the lowest-frequency sound. If they are closer, they are effectively coupled through the elasticity of the air.

At low frequencies, therefore, the two leaves forming the partition move together as a single wall, and the transmission loss is approximately equal to that of a single wall of the same total mass and stiffness. As the frequency rises, a resonance is encountered in which the two leaves move in antiphase, the stiffness of the air in the intervening cavity being added to that of the leaves. At still higher frequencies, air resonances are established in the cavity. All these modes of oscillation produce dips in the transmission curve, but all except the lowest of them can be reduced in depth by inserting into the cavity a blanket of mineral wool or porous foam with good sound-absorbing properties. Such treatment is almost invariably successful in increasing the transmission loss at middle and high frequencies, by damping natural oscillations.

The prediction of the transmission loss of even simple partitions is difficult by reason of the numerous factors entering it, and indeed, the exposition of these factors above may appear to have introduced inconsistencies between the general statements relating to those factors. The prediction of the behaviour of double partitions is extremely complicated, and cannot be said to have been solved even approximately. This applies with even more force to partitions with more than two leaves. The complication arises from the multidimensional nature of the phenomenon; the transmission loss is affected by the mass, stiffness, thickness, internal damping, and dimensions, of the partition. It is also necessary to consider bending waves, modal frequencies, the cavity depth and damping and, of course, the frequency. The loss is also vitally affected by the constraints on the edges of the material, e.g. whether it is firmly clamped at the edges or resiliently mounted.

It is not surprising that the best way to discover the exact sound-reduction index of a partition is still to measure it.

Some promising work has been done recently by teams at the universities of Liverpool and Salford, notably by Ford, Lord and Walker (1967), Ford, Lord and Williams (1967), and Mulholland, Parbrook and Cummings (1967), but the details of this work are outside the scope of this monograph. One difficulty is that transmission takes place as a result of interaction between the extremely complex modal systems of the partition and of the air in the rooms on either side of it. Classical methods of calculation are mainly confined to nonresonant regions of the spectrum and to comparatively simple resonant systems.

4.3.4   Power-flow method for transmission loss

An interesting statistical method that has been applied with success to airpanel systems at middle and high frequencies, where resonances are closely packed, is often referred to as the 'power-flow' method. This was first developed by Lyon (1964), Lyon and Eichler (1964), and others in connection with the excitation of vibrations in the structures of aircraft and space vehicles by airflow. Their methods have been simplified by Crocker and Price (1969) and applied with considerable success to the calculation of transmission loss through multiple partitions and cavity walls. Briefly, the method depends on the supposition that energy can be transferred from any mode in one system to a single mode in a second system coupled to it, and that the rate of transfer of energy between the modes is proportional to the difference between the energies in the two modes. This is analogous to the flow of heat between two bodies, which takes place at a rate proportional to the difference of their temperatures (or thermal-energy levels). Using the principle of the equipartition of energy, the excitation of any mode may be found by dividing the total energy in the system within a finite bandwidth by the number of modes in the bandwidth.

This principle will be illustrated by the simplest example, that of the coupling between the air in a room and a panel suspended in it.

Using the Crocker and Price notation, assume that the room and the panel are represented by the two boxes in Fig. 4.10.

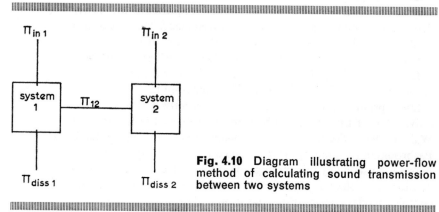

Fig. 4.10 Diagram illustrating power-flow method of calculating sound transmission between two systems

The arrows represent the flow of power from one component of the system to another, or of power in or out of the system. Π_{in1} and Π_{in2} represent power fed directly into the room and the panel, respectively, by loudspeaker or other transducer. Π_{diss} represents power dissipated in friction, and Π_{12} is the power flow from room to panel.

The equations of power flow are

$$\Pi_{in1} = \Pi_{diss1} + \Pi_{12} \quad \cdot \quad \cdot \quad \cdot \quad \cdot \quad (4.14)$$

$$\Pi_{in2} = \Pi_{diss2} - \Pi_{12} \quad \cdot \quad \cdot \quad \cdot \quad \cdot \quad (4.15)$$

We now introduce loss factors η_1, η_2, η_{12}, which we will define for the present purpose by the equation

$$\eta = R/M \quad \cdot \quad \cdot \quad \cdot \quad \cdot \quad \cdot \quad (4.16)$$

where R is the resistance, ω the circular frequency and M the mass associated with the mode in question. This is analogous to the electrical engineer's power factor, so that we may write

$$\Pi_{diss} = \eta E \quad \cdot \quad \cdot \quad \cdot \quad \cdot \quad \cdot \quad (4.17)$$

where E is the energy of the mode (i.e. the maximum kinetic or potential energy contained in the component when executing that mode of vibration).

Hence

$$\Pi_{12} = \eta_{12} n_1 (E_1/n_1 - E_2/n_2) \quad \cdot \quad \cdot \quad \cdot \quad (4.18)$$

where n_1 and n_2 are the number of modes per hertz at a frequency and E_1, E_2 are the modal energies at that frequency in the two components.

Further,

$$\Pi_{diss1} = n_1 E_1 \quad \cdot \quad \cdot \quad \cdot \quad \cdot \quad \cdot \quad (4.19)$$

From eqn. 4.16, it will be seen that the η values are directly related to the decay time of the room in the absence of the panel and to that of the panel alone, and may thus be calculated from simple measurements. n_1 and n_2, the modal densities can be calculated or found from steady-state measurements, and, finally, the input powers may be derived from the input velocity amplitudes and the input resistances.

With the aid of these parameters, the transmission-loss characteristic of the system can be found by solving the network equations.

This procedure can be extended to any number of components, thus allowing the transmission loss characteristic of complicated systems to be calculated from a knowledge of the separated components.

It will be clear that this method is suitable only for frequencies at which the resonant modes in all the components are densely packed. For regions of discrete resonances, classical methods must be used, but, with this restriction, the power-flow method offers great potential advantages in its application to components of shapes that preclude analytical treatment.

4.4   Practical design of sound-insulating partitions

4.4.1   General

From the foregoing brief review of the theory of sound transmission, it follows that the high sound reduction needed between a studio and its neighbouring areas requires the use of fairly heavy walls, and in most cases two or more separate leaves, and careful designing.

Before describing particular designs of sound-insulating partition, we will examine the effects of nonuniformity across the area of the partition, since it is not possible to have a studio entirely enclosed by uniform walls free from doors, observation windows or other features that produce variations in the local sound transmission.

If τ_1, τ_2,..., are the transmission coefficients of areas S_1, S_2,..., that together make up the whole area of a boundary wall, on the other side of which is a source of sound, it follows from the definition of transmission coefficient that the coefficient for the whole wall will be

$$\frac{S_1\tau_1 + S_2\tau_2 + ...}{S_1 + S_2 + ...} \quad \cdot \quad \cdot \quad \cdot \quad \cdot \quad (4.20)$$

Any small areas having very low values of τ will have a serious effect on the overall transmission loss of the partition, and this applies particularly to cracks, holes, service ducts and bad door seals. In any event, all the major components of each boundary of a studio, such as doors or observation windows, should be included in the summation, and precautions taken to avoid any direct airpaths, however small. Smith* has concluded from measurements on partitions pierced with circular holes and narrow slits that the effect of small airpaths of this kind can be fairly accurately predicted by assuming unity transmission coefficient for their exposed area, though, on the whole, such predictions tended to give slightly too low a value for the transmitted energy. All such apertures should be carefully sealed, since an open passage of typically 100mm² area will increase the sound transmission between two small studios by 1dB.

4.4.2 Insulating walls between technical areas in studio centres

For simplicity in discussing practical sound-insulating partition designs, the performance of a partition will be represented by the mean sound-reduction index, as defined in Section 4.1.2. For more detailed information on the constructions described, particularly in relation to the requirements shown in Fig. 4.2, it would be advisable to refer to Tables of sound-reduction index published by various testing authorities, or to carry out direct measurements.

The highest sound-level reductions required for studio walls or

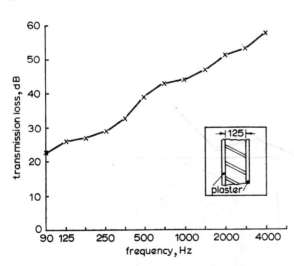

Fig. 4.11 Transmission-loss characteristic of whole-brick wall (plastered)

---

* SMITH, T. J. B. (1966): 'Sound insulation tests on lightweight camden-type partitions'. BBC Research Department internal note (unpublished).

roofs are of the order of 70 dB. A mean reduction of this amount would be required for the roof of a television or radio drama studio for protection against aircraft noise, or between a radio drama studio and one used for orchestral music. The lowest insulation normally specified for the walls of a studio is 45 dB. Typical designs to achieve mean sound reduction indexes between these figures will now be considered in successive steps of 5 dB.

45 dB mean s.r.i.

A 120 mm brick wall (halfbrick wall), plastered on both sides, will give a mean value of 45 dB. In practice, with the normal possibilities of flanking transmission provided by a single-leaf construction, this is not usually attained, and it is best to use a fullbrick wall (240 mm). The mass of the fullbrick wall, plastered on both sides, amounts to 450 kg/m². A typical transmission-loss characteristic is shown in Fig. 4.11.

Lighter walls to give the same reduction may be made from dry constructions with several leaves. A partition that is extremely efficient in relation to its mass and is consequently used extensively in BBC studio centres is named 'camden' partitioning, after the

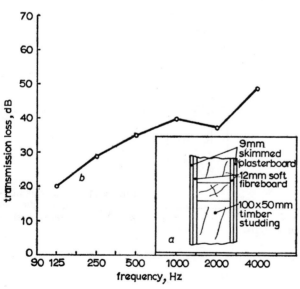

Fig. 4.12 Transmission-loss characteristic of 'camden walling'

studio centre in which it was first applied. Its construction and performance are shown in Fig. 4.12. It consists of a timber studding on each side of which is nailed a sheet of 12 mm wood fibreboard covered in turn by a sheet of 9 mm plasterboard that is finished with a skim coat of plaster. The mass is 28 kg/m², and the mean s.r.i. approximately 36 dB. The soft fibreboard contributes about 12% of the total mass, but its principal function appears to be that of isolating the plasterboard from the wooden frame and damping its resonances.

Two camden partitions together, with a 50 mm cavity between them, give a transmission loss, in practical situations, of 45 dB with a mass of 57 kg/m², which is only 13% of the mass of the brick wall with a similar mean transmission loss. Owing to the coincidence phenomenon, the curve of loss against frequency flattens out noticeably above 3 kHz, but this is of little consequence since the spectral content of most programme materials is falling rapidly at such frequencies.

A great advantage of a dry construction of this type is that it is easy to keep the cavity clear of any material that would cause bridging of the cavity. Moreover, the separate leaves are stable, and need no interconnecting ties such as those necessary between the leaves of a cavity-brick wall.

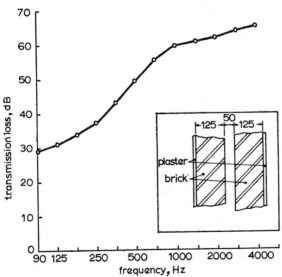

Fig. 4.13 Transmission-loss characteristic of plastered cavity-brick wall

Lighter constructions using more complicated combinations of materials to give the same mean transmission loss can be devised, but tend to become rapidly more costly as the complication increases. As an example, the transoms over the observation windows of the BBC Television Centre studios consist of alternate layers of thin steel sheet and felt impregnated with a polymer of high internal loss. There is a central cavity. The mean s.r.i. is 45dB, and the mass 25 kg/m².

50 dB mean s.r.i.

Two simple constructions capable of giving 50dB mean s.r.i. in practice are shown in Figs. 4.13 and 4.14. The wall consisting of two leaves of 120mm brickwork separated by a 50mm cavity has a slightly higher insulation value at low frequencies than that in which one of the leaves is replaced by a camden partition, but the latter weighs only 56% of the all-brick version.

55 dB mean s.r.i.

To produce a transmission loss of 55dB or more, a triple-wall structure is normally required, with considerable attention to the

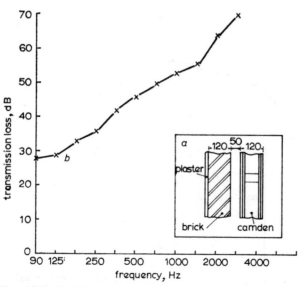

Fig. 4.14 Partition giving similar transmission loss to cavity-brick wall (Fig. 4.13), but with less mass

elimination of flanking paths. Any observation windows in the partition must be triple-glazed.

Fig. 4.15 shows a practical form of such a wall. It consists of a a 230mm brick wall between two leaves of plastered clinker block 80mm thick separated from it by cavities of 50mm. The mass is approximately 600 kg/m² and the mean s.r.i. 55dB.

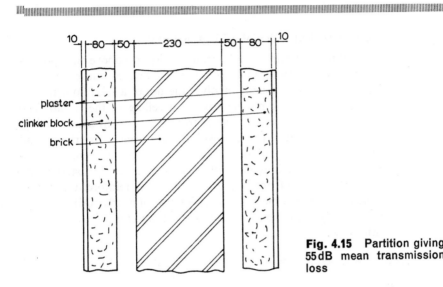

Fig. 4.15 Partition giving 55dB mean transmission loss

60dB mean s.r.i.

An appreciably higher transmission loss can be obtained from a 150mm clinker-block leaf and two 120mm leaves of brick, totalling 850kg/m² in mass. The clinker block used in these designs is porous, and absorbs sound well over a wide range of frequency, thus effectively damping cavity resonances. As this material is porous, it is essential that its outer surfaces be plastered or otherwise finished with an integral impermeable layer (Fig. 4.16).

65dB mean s.r.i.

A transmission loss of 65dB may be achieved by the use of a camden partition replacing half the thickness of the clinker in the design for 60dB given above. Fig. 4.17 shows the contruction and the measured characteristic.

These very high values of insulation can only be achieved in unbroken areas of wall; to avoid serious reduction of the overall transmission loss between the two sides, doors or observation windows breaking the continuity of the wall should be avoided.

70 dB mean s.r.i.

Sound-reduction indexes exceeding 65 dB can be obtained only by the use of two leaves of heavy masonry separated by a considerable airgap.

4.4.3   Practical design of ceilings and roofs

A ceiling separating a studio from a noisy area above must possess satisfactory impact-sound isolation as well as airborne-sound insulation. Design usually involves the use of a false floor for the area above, floating on the structural ceiling of the studio. The structural ceiling may be screened from the studio by a false ceiling, above which are carried low-frequency-sound absorbers, and services such as ventilation ducts. From considerations of load-bearing capacity, it is normally necessary to use a load-bearing floor consisting of not less than 120 mm of reinforced concrete, which itself gives a mean transmission loss of nearly 50 dB. Ceilings are usually unbroken by any sort of aperture, intentional or otherwise.

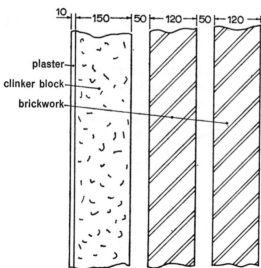

Fig. 4.16 Partition giving 60 dB mean transmission loss

All these contributing factors improve the sound reduction through a ceiling to a figure of 60 dB or more, and, except for a drama studio beneath an orchestral studio, no special provision is required for further airborne-sound insulation. In the two cases where this has occurred within the author's experience, two separate structural concrete skins, as described above, were used with complete success.

Roofs may require 60–70 dB mean sound insulation to provide adequate protection from the noise of aircraft, and the mass required to provide reductions in the upper part of this range in a single skin becomes economically and structurally unusable.

A sound-level reduction of 60 dB between the outside air and the floor of a large studio can be obtained by the use of a single layer of reinforced concrete topped with asphalt for protection against the weather; for higher figures, multileaf constructions must be used, and, once this decision is taken, it is comparatively cheap to build in a considerably increased transmission loss.

For example, a concrete slab 200 mm thick, giving a sound-level reduction of 55–60 dB at the floor of a studio, will need deep steel-work trusses to support it. The upper or lower faces of the truss can be designed economically to carry a timber roof also or a

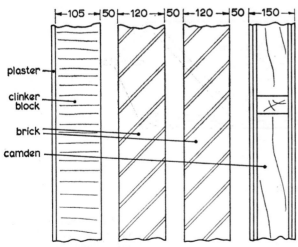

Fig. 4.17 Partition giving 65 dB mean transmission loss

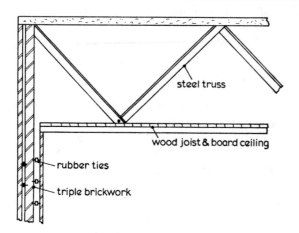

Fig. 4.18 Roof construction giving 70dB mean transmission loss

second concrete roof with an air space of 2–3m intervening in either case. These additional leaves, together with a cavity of 2–3 m, will raise the sound-level reduction at the floor to about 70dB and over 80dB, respectively. Figs. 4.18 and 4.19 show two such arrangements.

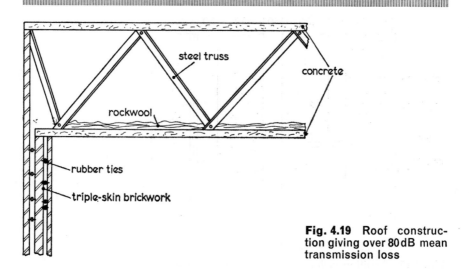

Fig. 4.19 Roof construction giving over 80dB mean transmission loss

4.4.4 Design of doors and windows

Observation windows in the walls separating studios from cubicles or recording rooms generally require two or three leaves of glass, which should not be all of the same thickness. Variation of the thickness prevents all the leaves having the same critical frequencies. Typical thicknesses are in the range 6–12mm. The reveals should be treated with packing felt or some similar sound-absorbing material to damp cavity modes, and arrangements should be made for opening the outer leaves for cleaning or removal of condensation, should this occur. They may be inclined, so that they do not cause disturbing reflections from the lights in the rooms. Such windows, however well designed, tend to be the weakest links in the wall, and their design and construction should be supervised with care to ensure that they provide as little sound transmission as possible.

Doors also need careful consideration. It will be clear that precautions must be taken to prevent cracks round the edges, the effect of which would be to impair the sound insulation. To obtain the greatest possible freedom from such gaps, the best way is to provide wedge-shaped door stiles and lever handles operating linkages by which the door can be forced under pressure into the frame, in the same manner as the doors of large meat-storage

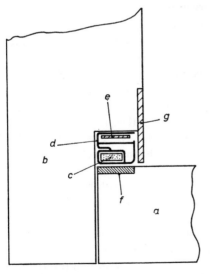

Fig. 4.20 Magnetic door seal for preventing leakage through cracks round frame band jambs

a Door
b Door frame
c Magnetic strip
d Plastic extrusion
e Nonmagnetic securing plate
f Mild-steel strip
g Cover strip

Courtesy: BBC

refrigerators. In the UK, however, the positive bolting of doors in this way is discouraged, partly because of the difficulty of escape in case of fire, and partly in the interests of easier access for production purposes. It is therefore necessary to use spring door closers and efficient edge seals. Quite good results can be obtained by the careful application of proprietary draught strips and raised thresholds, and much ingenuity has been exercised in the past in devising edge seals that operate satisfactorily with a small closing pressure and continue to function even when the door warps. All these may be said to be superseded by the use of magnetic-strip seals (Fig. 4.20).

A single door capable of being opened quickly and without undue effort is unlikely to give more than 35dB mean sound-transmission loss; doors must therefore always be used in pairs with a space or sound lobby in between. The lobby should be surfaced inside above dado level with efficient sound-absorbing materials to reduce the reverberation and hence the coupling between the two doors.

Other information on the design of studio details for good sound insulation has been given by Brown (1964).

4.5 Methods of measurement

4.5.1 British Standard measurements

BS2750 gives details of the recommended method of measuring the sound-level difference between two points. Sound is radiated into the source room from one or more loudspeakers, and the sound-pressure level is measured at several points in each room in turn, using a square-law meter. The s.p.l. corresponding to the mean intensity at the measuring points is then calculated by converting the measured levels into power ratios, taking the average and reconverting to sound-pressure levels. The power ratios corresponding to decibels above a reference level are given in Table 2.1.

The mean s.p.l. in the receiving room is then subtracted from that in the source room at each frequency, and a curve of level reduction is plotted.

It is recommended that the positions for measurement be chosen at distances more than half a wavelength from any wall. Waterhouse (1955) has shown that the pressure close to a wall in a reverberant room is 3dB higher than that in the middle of the room; close to an edge it is 6dB higher, and close to a corner it is 9dB higher, than in positions well away from walls.

Thus sound-pressure levels will be, on the average, 3dB higher over the central parts of the walls than in the body of the room, and this must be taken into account if microphone positions other than the recommended ones are used.

4.5.2 Other methods of measurement

The simple method of measurement described above gives only the reduction of sound level between the two points, and can provide no information about the paths of transmission or their relative importance. For all remedial work, it is necessary to have such information, and supplementary tests have been devised for the purpose. These will be described in Chapters 5 and 6.

Chapter 5

Structure-borne sound

5.1 Propagation of sound through solids

5.1.1 General

Structure-borne sound transmission in a building is conventionally taken to refer to the transmission of sound from a vibrating body in contact with the structure of the building. Thereafter, its behaviour is identical with that of 'airborne' sound, i.e. that part of the sound energy which is first impressed on the structure by wave motion in the air surrounding the source.

Thus the treatment of sound insulation in Chapter 4 was confined to the examination of the transmission of sound between adjacent spaces separated by a common partition or wall. The sound pressures acting here on one side of the dividing wall set the whole of it into motion, and sound waves are radiated from its further face as a result.

The vibrations of the partition may, however, be transmitted into the structure of the building generally, and sound may consequently be radiated into rooms far from the original source of sound, because the attenuation of sound in passing through solid building materials is comparatively slight. This is a very important feature of steel-framed buildings in particular, and the consequences will be described below.

Propagation may take place in different ways:

(a) as pure extensional waves, in which the particle velocities are entirely in the direction of propagation

(b) in structural elements of which the cross-sectional dimensions are small compared with the wavelength, as Young's-modulus waves in which displacements along the direction of propagation are accompanied by Poisson expansions and contractions at right angles to the direction of propagation

(c)   as shear waves, in which the particles vibrate at right angles to the direction of propagation

(d)   as bending waves, normally in materials of relatively small transverse dimensions, where a curvature is propagated along the material

(e)   as torsional waves, in which the elements of the material rotate about the neutral axis of the section

(f)   as Rayleigh waves, in which only the surface of the material and the adjacent layers are deformed.

Fig. 5.1 illustrates these six types of wave-motion.

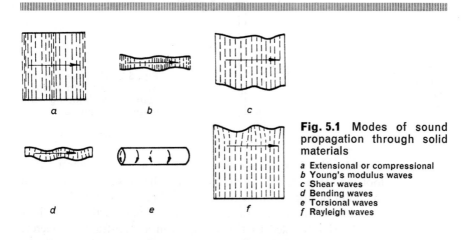

Fig. 5.1 Modes of sound propagation through solid materials

a Extensional or compressional
b Young's modulus waves
c Shear waves
d Bending waves
e Torsional waves
f Rayleigh waves

If we examine the first two types in the list above, we see that the appropriate elastic moduli for calculating the velocities of propagation are the bulk modulus and the Young's modulus, respectively. The bulk modulus is always higher than the Young's modulus, and is often of a completely different order. For instance, in rubber, pure extensional waves travel at 1500 m/s, whereas Young's-modulus waves in thin threads have a velocity of only about 50 m/s. In steel, bulk-modulus waves travel at 5000 m/s and, therefore, even at frequencies as high as 5 kHz, the wavelength will be greater than the thickness of any steel building element.

Torsional waves are of great importance in some fields of study, such as vibration in road vehicles, but they are seldom excited to any significant extent in buildings. Rayleigh waves could occur at high frequencies in the surfaces of very thick walls, but would be so exceptional as to have no real importance.

Thus we are left with two types of wave motion: Young's-modulus waves and bending waves. These two types will be interchanged during transmission through a building, as the following example will show:

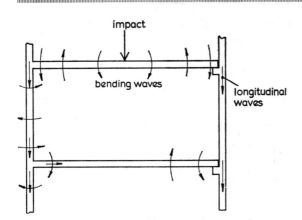

impact

bending waves

longitudinal
waves

Fig. 5.2 Propagation of waves from an impact source as bending and longitudinal motion

Consider the floor of a room excited with vertical vibration by footsteps (Fig. 5.2). The vibration will be travelling away from the point of impact in the form of bending waves until they reach the walls of the room. In a conventional construction, the floor may be rigidly fixed to the walls, as in a steel-framed concrete structure, or carried on bearers, as in an ordinary house. In the first, the wall will be excited into both bending and longitudinal waves; in the second, the bending waves will be partly reflected and partly converted into longitudinal waves in the walls.

On meeting the lower floor, the longitudinal waves will be converted into bending waves, and the bending waves will remain as bending waves or be partly converted into longitudinal waves, according to the stiffness of the junction.

5.1.2   Analysis of waves in solids

The properties of waves transmitted as compression or bending of the material are summarised below. A fuller account has been given by Cremer (1953) and by Cremer and Heckl (1967), whose notation will be used here.

In general, we are interested only in the propagation of these two wave types in rods and plates of which the transverse dimensions are small compared with the wavelength of the sound at audio frequencies and with their lengths in the direction of propagation.

Longitudinal waves

The velocity of waves in a thin rod is given by

$$c_L = \sqrt{(E/\rho)} \quad . \quad . \quad . \quad . \quad . \quad (5.1)$$

where E is the Young's modulus and ρ the density, but, in a thin plate in which lateral contraction can take place only perpendicular to the plane of the plate, this is replaced by a larger value

$$\sqrt{\{E/\rho(1-\mu^2)\}} \quad . \quad . \quad . \quad . \quad (5.2)$$

where μ is the Poisson's ratio for the material.

The wavelengths are, respectively,

$$\lambda_L = c_L/f \text{ for a rod} \quad . \quad . \quad . \quad . \quad (5.3)$$

and

$$c_L/f(1-\mu^2)^{\frac{1}{2}} \text{ for a plate} \quad . \quad . \quad . \quad (5.4)$$

However, since μ is necessarily less than $\frac{1}{2}$ (the value for an incompressible liquid), and normally very much lower than $\frac{1}{2}$ for most building materials, the factor $(1-\mu)^{\frac{1}{2}}$ lies between 0·87 and 1, so that, for most purposes, it may be taken as unity, having regard to the uncertainties due to variations in the moduli of building materials.

The longitudinal impedance of a rod, by analogy with that of a column of air, is

$$Z_L = \sqrt{(EA^2\rho)} \quad . \quad . \quad . \quad . \quad (5.5)$$

where A is the cross-sectional area.

Bending waves

The bending waves in rods and plates again travel at similar speeds, which depend on both the thickness of the material and the frequency. This was noted in Chapter 4, in connection with the coincidence effect in partitions. For plates or rectangular-section rods, the velocity of bending waves is given by

$$c_B = \sqrt{(1\cdot8c_L hf)} \quad . \quad . \quad . \quad . \quad (5.6)$$

where h is the thickness and

D

$$\lambda_B = \sqrt{(1{\cdot}8c_L h/f)} \quad . \quad . \quad . \quad . \quad (5.7)$$

is the wavelength.

It will be seen that, for a given material, the velocity is proportional to the square root of the thickness and the square root of the frequency. The wavelength is inversely proportional to the square root of the frequency.

The effective impedance presented by the end of a bar in transverse motion is

$$Z_B = \sqrt{(mB)} \quad . \quad . \quad . \quad . \quad (5.8)$$

where m is the mass per unit length, and B is the stiffness given by EI where I is the second moment of area (often called the 'moment of inertia') of the section about the neutral axis.

For a long rod, the midpoint transverse impedance is

$$4\sqrt{(mB)} \quad . \quad . \quad . \quad . \quad . \quad (5.9)$$

and, for a point in a plate of infinite area, the impedance to transverse deflection is

$$\frac{4}{\sqrt{3}} c_L \sigma h^2 \quad . \quad . \quad . \quad . \quad (5.10)$$

where σ is the mass/unit area. Unlike the wave velocity, this is independent of frequency.

Discontinuities

The impedance values given above enable one to calculate the attenuations caused by junctions between contingent building elements, remembering that there will be an interchange of energy between the two types of wave motion wherever there is a change of direction.

Changes of section produce only small attenuations, e.g. an attenuation of 3 dB in passing through a change of section of approximately 5:1. These attenuations are rather greater for bending waves than for longitudinal waves.

A mass fixed rigidly to a rod is more effective than a simple discontinuity in producing attenuations, the attenuation increasing by 6 dB for every doubling of the frequency. This is analogous to the behaviour of a simple massive partition towards normally incident airwaves.

The most effective way of producing attenuations in a building element is, however, to introduce a resilient layer into the path of the sound waves. The effects of such layers can be calculated quite accurately from a knowledge of the densities and elasticities of the materials. The extreme example of this method is the introduction of an airgap into the conducting element.

The attenuation given by such a layer may be calculated (Cremer, 1953, p. 329) by the expression

$$R = 20 \log_{10} \frac{\omega t Z_1}{2E_2} \qquad . \quad . \quad . \quad . \quad (5.11)$$

where Z_1 is the impedance of the interrupted structural member, t and E_2 are the thickness and Young's modulus of the elastic layer, and $\dfrac{\omega}{2\pi}$ is the frequency.

Practical embodiments of this principle are given in Section 5.3.

5.2 Significance of structure-borne sound in buildings

It is necessary to understand the behaviour of structure-borne sound in some depth for three reasons. First, impulsive sounds exciting the solid structure of a building are propagated with comparatively little attenuation, and thus affect quiet areas far from the source of sound. In a steel-framed building, in which there are few discontinuities to impede transmission, it is possible to hear impulsive sounds of all kinds coming from other parts of the building. In a quiet room, one can hear the sounds of footsteps, electric-light switches, heating pumps, the banging of doors, and sounds of plumbing, lift motors, scraping of furniture over floors, and so on. These noises are often very obvious in blocks of flats, even though the airborne-sound insulation between the flats may be adequate. If these sounds were to be allowed to penetrate a studio, the effects on a broadcast would be quite intolerable.

Secondly, as mentioned in Chapter 4, the airborne-sound transmission between two rooms may be supplemented by energy travelling along a number of solid paths parallel to the direct path. The transmission by these paths places an upper limit on the sound reduction that may be obtained by improvements to the partition directly separating them.

Thirdly, for the reason just mentioned, the possibilities of sound transmission through the structure must receive the first consideration in the design of a studio centre; the transmission losses of the internal partitions can be decided at a later stage, or adjusted to different values, but only in so far as flanking transmission inherent in the design of the building as a whole is low enough. For example, if there are two adjacent rooms formed from walls of 150 mm concrete, with floors and ceilings of a similar material, it is not possible to improve the sound reduction from one to the other simply by adding another leaf to the partition wall, because approximately half the sound energy already passes between the rooms by way of indirect paths, such as those shown in Fig. 4.1.

The quantitative relationship between airborne and structure-borne transmission may now be understood by reference to Fig. 5.3.

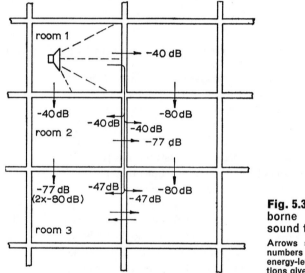

Fig. 5.3 Propagation of airborne and structure-borne sound through large building

Arrows show paths of propagation, numbers represent approximate energy-level contributions, on assumptions given in text

Assume that there is a source of airborne sound in room 1. This sets the structure into vibrations of, say, such amplitude that the pressure level of the sound radiated from the ceiling of room 2 is 40 dB lower than that in the first room. The vibrations simultaneously induced in the flanking walls will be of the same order of magnitude as those causing the direct sound transmission. Before reaching the flanking walls of room 2, however, these vibrations will have been attenuated by about 7 dB as a result of internal losses, reflection from the junction of the walls with the floor, and spreading out into the other rooms of the same storey.

The sound radiated into room 2 by the four walls will be 6dB greater in power than that from one wall, and, if the ceiling is of about the same area as of a single wall, the total radiation from the four walls will thus be about the same as from the ceiling. This is in agreement with the observation (Meyer *et al.*, 1951) that the flanking transmission in a uniform structure is of the same order of magnitude as the direct transmission between adjacent rooms.

If we follow the structure-borne sound down to the next storey, we find that it has suffered a further attenuation of 7dB, but the sound coming directly through the floor of room 2 has been attenuated by a further 40dB. We conclude, therefore, that, in the absence of the flanking sound, the sound-level reduction would have been 80dB, but that the flanking component has been attenuated by only 47dB. In room 3, therefore, the greatest part of the sound energy has arrived by paths through the flanking structure. Following this process to other rooms in the building, we find that, the further we recede from the source of sound, the more important, relatively, do the flanking paths become.

Consequently, once vibrational energy is present in the structure of a building, every area in which a low noise level is to be maintained must be carefully isolated from any solid contact with the structure. It is more satisfactory to design the building from the start, so that all possible sources of impulsive sounds are themselves isolated from the structure. There are at present no satisfactory means of substantially reducing the transmission of sound through the main structure of a building, though this is the subject of a good deal of research. It has been remarked that effective isolation can be obtained by the use of resilient layers, but these are generally not compatible with the structural strength required in the main frame of a large building. The most practicable solution is usually for both sound sources and sensitive areas to be on resiliently mounted subsidiary frames, as will be described below.

5.3 Design for isolation of structure-borne sound

5.3.1 Protection from vibration and sound external to building

It has been suggested above that reduction of structure-borne sound should be carried out, if possible, at the source. This clearly cannot be done where the source is not under the control of the studio designer, as, for example, traffic or railway noise.

Several studio centres in the London area are situated over, or very close to, underground railway lines, and present considerable problems in vibration isolation. The most interesting example is that of the Broadcasting House extension in Great Portland Street, London. In this building there are studios as close as 10 m from the Bakerloo southbound railway. When the shell of the building was first completed, the sound level in the interior during the passage of a train reached over 75 dBA. One method of achieving the necessary level reductions would have been to mount the whole building on resilient supports; this form of isolation was used for a block of flats built over the St. James's Park underground station (Waller, 1966). However, this would have been uneconomical for a studio centre of the projected type, since, for reasons unconnected with technical requirements, there were to be no noise-sensitive broadcasting areas above the ground floor.

It was therefore decided to isolate the sensitive areas individually, a method that has been used generally both in the UK and abroad.

5.3.2 Theory of isolation of studios by resilient mountings

Fig. 5.4a is a diagram of a studio simply mounted on a pair of springs of total stiffness factor (the ratio of load to deflection) k. The resistive losses in the deflection of the spring are assumed to be viscous, and of a magnitude Ri, where i is the vertical velocity of the studio on the suspension and R is the resistance constant. This resistance is conventionally represented by the dashpot symbol. We assume that the ground on which the springs stand is oscillating vertically at a frequency $\omega/2\pi$ and that its mass is so large that these vibrations are unaffected by any consequent motion (assumed to be also vertical) of the studio on the springs. Fig. 5.4b is the corresponding electrical analogue of the mechanical system. I is the constant-amplitude velocity of the floor, and i the resulting velocity of the studio. R represents the frictional element, and M

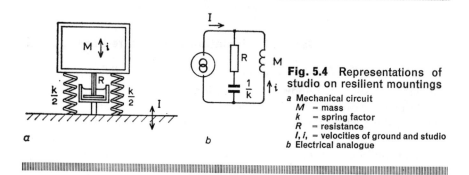

Fig. 5.4 Representations of studio on resilient mountings
a Mechanical circuit
  M = mass
  k = spring factor
  R = resistance
  $I, i,$ = velocities of ground and studio
b Electrical analogue

the mass of the studio, and the capacity representing the spring compliance is $1/k$.

We have to find the value of $|i|/|I|$, which represents the ratio of the velocity amplitude of the studio to that of the ground.

Using Kirchhoff's network laws, we have, for the mesh on the right,

$$j\omega Mi - (I-i)(R+k/j\omega) = 0 \quad . \quad . \quad . \quad (5.12)$$

whence

$$i/I = \frac{R+k/j\omega}{R+j\omega M+k/j\omega}$$

Since I is a constant exciting current, the quadrature components in the two branches must be equal and opposite, so that

$$\frac{|i|}{|I|} = \frac{|R+k/j\omega|}{|R+j(\omega M-k/\omega)|} \quad . \quad . \quad . \quad (5.13)$$

$$= \sqrt{\left\{ \frac{R^2\omega^2/k^2+1}{R^2\omega^2/k^2+(\omega^2 M/k-1)^2} \right\}} \quad . \quad . \quad . \quad (5.14)$$

Writing M/k as ω_0 and the ratio $|i|/|I|$ as τ (a transmission coefficient analogous to that defined in Chapter 4 for airborne-sound transmission), we have

$$\tau = \sqrt{\left\{ \frac{R^2\omega^2/k^2+1}{R^2\omega^2/k^2+(\omega^2/\omega_0^2-1)^2} \right\}}$$

We may reduce this equation to a nondimensional form if we write $Q = M\omega_0/R$, where Q is seen to be analogous to the Q factor of a resonant circuit, and $\beta = \omega/\omega_0$:

$$\tau = \sqrt{\left\{ \frac{1+\beta^2/Q^2}{(\beta^2-1)^2+\beta^2/Q^2} \right\}} \quad . \quad . \quad . \quad (5.15)$$

If Q is large, as in a lightly damped oscillation,

$$\tau = \sqrt{\left(\frac{1}{(\beta^2-1)^2} \right)}$$

which is clearly a maximum for $\beta = 1$. ω_0 is, therefore (as the reader will already have guessed), the frequency of resonance.

If $\beta = 1$, we have $\tau = \sqrt{(1+Q^2)} \simeq Q$

If $\tau = 1$, moreover, $\beta = 0$ or $\sqrt{2}$, independently of the value of Q.

Thus the curves of transmission coefficient diverge from a common point at zero frequency and cross again at a frequency $\sqrt{2}\omega_0$. Fig. 5.5 shows curves of $-20 \log_{10}\tau$ plotted against ω/ω_0 for representative values of Q. This parameter corresponds to the transmission loss, in decibels, of a partition.

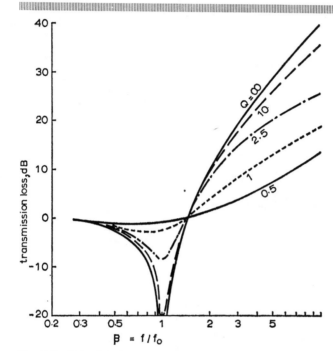

Fig. 5.5 Calculated transmission loss through spring mounting
Parameter: Q factor of mounting

Note that, at frequencies below $\sqrt{2}\omega_0$, this is negative. For this reason, the natural frequency of a studio mounting should be chosen to be well below the lowest frequency of vibration against which protection is required. Above $\sqrt{2}\omega_0$, the higher the value of Q, the higher the transmission loss. Thus it is best to use mountings with little damping if the natural frequency can be kept low enough. However, a highly damped mounting may give a useful degree of isolation even if its natural frequency is within the working range, because the additional transmission at resonance can be made small compared with the improvement over the rest of the frequency compass.

5.3.3 Practical designs of studio suspension

The last sentence above applies particularly to the commonest form of suspension used for studios, and, indeed in many other buildings. This consists of a blanket of mineral wool of about 25 mm uncompressed thickness, laid on the structural floor and carrying a cement screed, which forms the floor of the room. A typical arrangement is shown in Fig. 5.6. Before screeding, the mineral wool is covered with a layer of waterproof building paper to prevent leakage of the cement through the wool. The walls of the room may be built on a strip of cork and isolated from the floor by packing felt. The weight of the screed is usually sufficient to compress the mineral wool, so that its spring factor is raised and yields a natural frequency of about 100 Hz. However, this is combined with a fairly low value of Q, and, as explained above, it thus gives protection against many structure-borne sounds. It is generally satisfactory for speech studios, unless there is a very high level of low-frequency vibration, because the lower speech frequencies are often attenuated for other reasons by means of electrical filters.

Alternatively, instead of the cement screed, the floor may consist of wooden boards laid on battens that are carried by sleeper joists

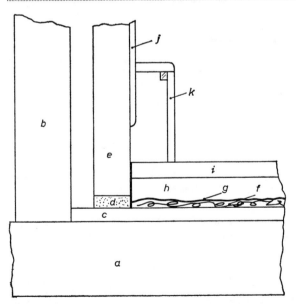

Fig. 5.6 Studio floor laid on mineral wool

a Structural floor slab
b 125 mm brickwork
c Lightweight concrete screed
d Cork
e 75 mm breeze block
f Mineral-wool blanket
g Waterproof paper
h Lightweight concrete
i Finishing screed
j Plaster
k Microphone skirting

laid on the fibre mat. A disadvantage is that microphone floor-stands cannot be used on the floor, because it is flexible and readily transmits vibrations from footsteps and other movements to the microphone.

For protection against high levels of interference, it is better to use steel- or rubber-spring systems having lower natural frequencies and higher Q values than can be obtained with mineral-wool blankets. However, for such spring suspensions, the studio floor must have considerable inherent rigidity, since it will be carried only over a comparatively small fraction of its area.

Steel springs are favoured in Europe, and particularly in Germany. The studios in the Munich Centre, built in 1959, are hung from vertical tie rods that are attached to helical compression springs mounted on brackets on the outer walls (Struve, 1964). The commonest method, however, is to use a steel I beam to form the periphery of the floor; the beam stands on a continuous corrugated strip of spring steel or a number of C-shaped steel springs fixed to the ground. Fig. 5.7 shows these two arrangements. Permanent jacks may be built into the system, by which the floor can be raised for replacing the springs, should this ever be necessary.

a   *b*

Fig. 5.7 Two forms of steel spring for studio suspension
a C-type
b Continuous strip

The floor mounting used by the BBC in the extension to Broadcasting House mentioned above has been described by Brown (1964). Rubber machine-mounting pads were chosen as the spring elements. These consist of flat sheets of rubber approximately 300mm square, on one side of which are moulded zigzag ribs, which carry the load. The zigzag form ensures stability under high deflection.

The pads were laid round the periphery of the studio space with intermediate lines of pads running in one direction. These were then spanned by prestressed-concrete blanks to form a basefloor, around the edges of which the brick walls were built. Finally, a screed was added to complete the floor, and a ceiling was constructed over the walls. The studio thus consisted of a complete box mounted on the rubber pads (Fig. 5.8).

The expected reduction of vibration at 100 Hz was 20 dB, but, in the event, a reduction of only about 10 dB was obtained. This was fortunately enough to give the required low background-noise level, but the discrepancy was disturbing enough to warrant investigation. Ward and Randall (1966) found it to be a result of an abnormally high Q factor of the slab in bending vibration, which greatly increased the coupling in the airspace. They found that the addition of a layer of cement-asphalt mix of high internal loss before the application of the screed resulted in a great improve-

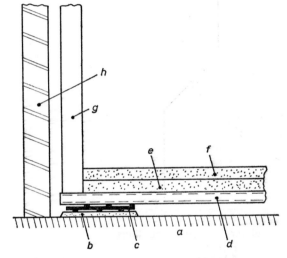

Fig. 5.8 Suspension of studio on rubber pads

a Structural floor
b Concrete plinth
c Rubber pad
d Prestressed-concrete planks
e Concrete
f Screed of asphalt concrete
g Inner wall
h Outer wall

ment in the isolation. (Recent experiments by the author at the University of Aston have shown that there is a significant reduction in internal losses in prestressed concrete beams, compared with unstressed reinforced bars.)

No provision was made for changing the rubber pads, since the allowance of jacking clearances throughout the complex of approximately 60 studios and other technical areas would have greatly increased the volume and cost. Case histories of rubber-mounted rooms up to 30 years old indicated that no serious permanent setback need be anticipated, provided that the mountings were not exposed to solvents or ultraviolet light. On the other hand, the fact that rubber tends to stiffen slowly with age must be taken into consideration in design. Waller (1966) concluded that permanent creep amounted to only 10% of initial deflection in a decade.

An alternative type of rubber spring was used in the mounting of one studio at the Television Centre, London. These consisted of alternative layers of steel sheet and rubber (Fig. 5.9). The rubber is deflected in a combination of compression and shear, giving adequate compliance combined with great stability.

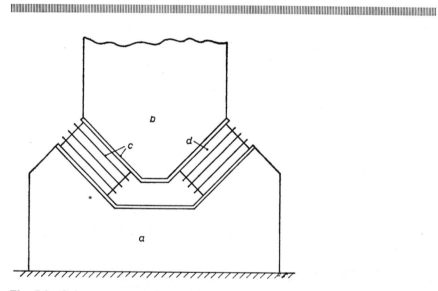

Fig. 5.9 Column supported on rubber-sandwich mountings
a Foundation   *c* Steel
b Concrete pillar   *d* Rubber

If reference is made again to Fig. 5.5, it will be seen that the transmission loss at any particular frequency depends only on the Q factor of the mounting and the ratio of the frequency to the natural frequency of the mounting. Increase of Q above 10, a factor that is attainable with rubber or steel springs, makes little difference to the transmission loss above $\sqrt{2}\omega_0$, and we may therefore assume this Q factor for design purposes. Knowing that the attenuation required at low frequencies, we may therefore decide on the natural frequency of the mounting.

Now, if we assume that the mounting behaves linearly, so that $\omega_0 = \sqrt{(k/M)}$, we may also write down the static deflection of the springs under the weight of the studio as Mg/k, which we will call d.

Then from these two equations, $\omega_0 = \sqrt{(g/d)}$, and, if we substitute the value of g and put $f_0 = \omega/2\pi$, we have, finally,

$$f_0 = 14\cdot3/\sqrt{d} \text{ hertz} \quad . \quad . \quad . \quad . \quad (5.16)$$

where d is in millimetres.

5.3.4 Measures to prevent transmission of impact noises into building structure

The previous Section was concerned with the prevention of incoming vibrations of which the sources were not accessible to control. Precautions should also be taken to prevent the excitation of the building structure by impacts originating in it. The commonest sources of such impacts are footsteps; electric-light switches; the moving of furniture, pianos etc.; and the banging of doors. Of these, footsteps are usually the most persistent.

There are two main methods of reducing the transmission of impulses to the building, both involving the principle of resilient connections. The simplest and cheapest method is to cover with carpet or a rubber flooring backed with sponge all floors from which sound is likely to be transmitted to quiet areas. The disadvantage of such a flooring is that it can easily be compressed by heavy impacts to a point where it ceases to provide the necessary compliance, and the isolation breaks down. Nevertheless, considerable reduction is obtained for normal footsteps, even if occasional heavy bumps get through.

A generally more satisfactory method is to mount the whole floor on a mineral-fibre mat, as described in Section 5.3.3. Such a comparatively expensive construction is justified for rooms immediately adjacent to studios, such as their own control cubicles.

5.3.5 Isolation of services

It is necessary also to take measures against the transmission of operating noises from various service equipment, such as power transformers, automatic voltage regulators, ventilating fans and motors, circulating pumps, taps, toilet flushes and lifts. The first five of these can be dealt with satisfactorily by means of antivibration mountings, as described in connection with voltage regulators and ventilation equipment in Chapter 3. They have the common feature that their noise spectrum is limited to a few discrete frequencies; transformers emit vibration at the supply frequency and some of its harmonics, whereas, for the motor-driven equipment operating at constant speed, the frequency of rotation is the determining factor.

A machine on an antivibration mounting may be represented by the mechanical circuit of Fig. 5.10*a* and its electrical analogue in Fig. 5.10*b*. The alternating forces produced by the machine are represented by the constant voltage E, and we wish to know the value of the ratio $|e|/|E|$ of the force across the mounting to that produced by the machine.

Now

$$|e|/|E| = \frac{|\text{impedance of } R \text{ and } 1/k|}{|\text{impedance of whole circuit}|} \qquad . \quad . \quad (5.17)$$

$$= \frac{|R+k/j|}{|R+jM+k/j|} \qquad . \quad . \quad . \quad . \quad (5.18)$$

This is identical with eqn. 5.13, and we may therefore derive the force ratio that we require from eqn. 5.15 and the velocity-ratio curves of Fig. 5.5.

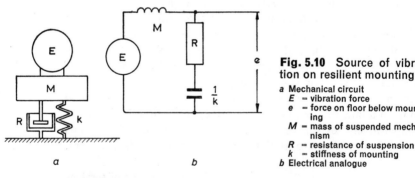

Fig. 5.10 Source of vibration on resilient mounting

a Mechanical circuit
E = vibration force
e = force on floor below mounting
M = mass of suspended mechanism
R = resistance of suspension
k = stiffness of mounting
b Electrical analogue

The design of antivibration mountings is actually a more skilled matter than this simple exposition would indicate, since all types of motion (linear and rotational) must receive attention, and the efficacy of the system is influenced by the stress/strain characteristic of the spring element. For more information, the reader should study a specialised work, such as that by Snowdon (1968).

When equipment is installed on antivibration mountings, all conduits, pipes and other links between the equipment and the building should be made flexible or in other ways prevented from coupling the equipment to it.

All the pipes of the plumbing services should be mounted on walls, where this is necessary, with clips incorporating rubber or soft-plastics bushes to prevent the transmission of noise from ball valves or other points where turbulence occurs in water flow. A frequent source of clicks is the pushbutton type of tap fitted in public washrooms to avoid waste of water. As the seatings of these taps approach closure point, the acceleration of the water through the orifice sets up a strong Venturi effect, which causes the seating to snap down with the production of impulsive sound and vibration. This effect can be reduced by good design and maintenance, but it is better not to install taps of this kind in studio centres. Toilet flushes vary greatly in the noise that they produce. In the commonest type of 'w.c.', the so-called 'washdown' closet, the action depends on creating a sudden release of the water at high velocity. This is usually accompanied by a high level of noise, and means should be found of preventing it from being communicated to the main strucutre. The base of the pan should be resiliently mounted and the soil-pipe connection should be made with a rubber-ring seal, as used with polypropylene soil pipes. Syphonic w.c.s are very much more silent, since their action does not depend on high water velocities. The small extra cost of this type is offset by the fact that it can be installed with little or no attention to noise transmission. Quiet ball valves, in which cavitation noise is eliminated (Sobolev, 1955), should be fitted.

Lifts should be sited as far as practicable from quiet areas, though, in the author's experience, there is seldom any noise nuisance from lifts designed with a view to silent operation.

Quite apart from the transmission of noises generated in the plumbing systems, pipes carrying water or other liquids can also conduct extraneous sounds into quiet areas. For instance, if a studio is heated by radiators supplied with hot water from an external boiler, the pipes and radiators can radiate sound as well as heat if, at some part of the circuit, the supply or return pipes pass through a noisy area or are exposed to casual impacts.

This can be satisfactorily prevented by the use of armoured rubber coupling pipes; it is found best to use two such connections, one on each side of a right-angle bend. They should be inserted outside, but as near as possible to, the studio. Fig. 5.11a shows the preferred arrangement of couplings on a pipe in a test situation in which a pipe passes through a wall to a radiator in a reverberation room (Jones, 1967*b*). Measurements were made of the sound level due to a loudspeaker outside the room and the level created within the room by the sound transmitted by the pipe and radiated by the by the pipe and radiator. Fig. 5.11*b* shows the difference between the sound levels in the room, with a rigid pipe and after the insertion of the couplings.

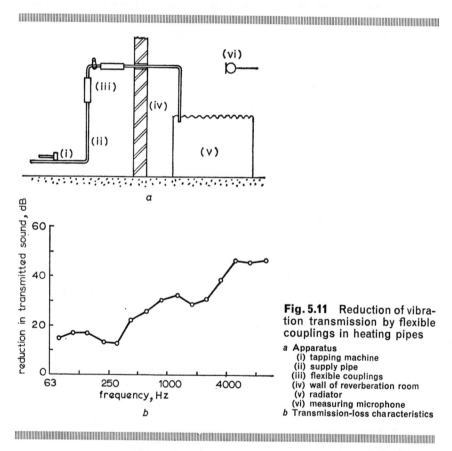

Fig. 5.11 Reduction of vibration transmission by flexible couplings in heating pipes

a Apparatus
    (i) tapping machine
    (ii) supply pipe
    (iii) flexible couplings
    (iv) wall of reverberation room
    (v) radiator
    (vi) measuring microphone
b Transmission-loss characteristics

5.3.6 Flanking transmission

The unwanted transmission of sound from a noisy to a quiet area by solid flanking paths has already been mentioned briefly, as one of the important features of structure-borne sound. Before any partition is designed, it is necessary to examine the probable transmission by such paths and to decide the effective transmission loss of the paths taken together. This figure has been called by

Gösele (1954) the 'maximum possible sound insulation' between the areas, because a partition added between them could not raise the transmission loss above this figure. As shown by Meyer *et al.* (1951), a brick or concrete partition in a building consisting mainly of single-leaf walls and floors of a similar material has an airborne-sound-reduction index similar in magnitude to that through all the flanking paths combined, and consequently the maximum attainable transmission between adjacent areas in such a building is about equal to that of a single-leaf wall. In situations where a greater loss is required for broadcast purposes, therefore, adequate measures must be taken to increase this maximum possible loss. The flanking transmission by each path must be dealt with to increase the maximum possible loss by the same amount as the desired increase in the total transmission loss.

Consider, for instance, a floor of an administrative area, below which is a studio. It will be necessary, first, to add a floating floor with the object of cutting down the footsteps, but any walls common to the upper and lower rooms will conduct sound into the lower. On the principle of preventing transmission of noise and impacts into the structure, it will be best to screen the walls of the upper room with a second leaf of a type that will raise its transmission loss by the required improvement in the maximum possible transmission loss. In most practical instances in broadcasting centres, a stud framing carrying a layer of plasterboard over one of soft fibreboard 12mm thick will provide the necessary reduction

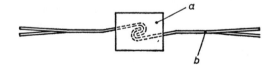

Fig. 5.12 Simple resilient wall tie. Note failsafe provision

a Soft rubber
b Mild-steel tangs for insertion in mortar joints

in transmission if the studding is maintained clear of the main wall. In others, a leaf of 75mm clinker block may be required. Again, this should be clear of the main wall and should stand on a plinth of packing felt or cork. It may be connected to the main wall for the sake of stability by means of wall ties, but these must be of a flexible type, such as those shown in Fig. 5.12.

If a higher degree of protection against flanking transmission is required, the lower room may be provided with a suspended ceiling and its walls enclosed behind additional leaves of suitable construction as for the upper room. Similar measures will be applicable to the rooms on the same floor as the source room.

Detailed attention is required for the design of doorways and window frames for the avoidance of transmission paths across wall cavities. A satisfactory design is to have a common door frame split in the cavity between the walls of the two communicating rooms. The gap is packed with felt and covered by a fillet that is fixed to only one of the frames.

5.4   Measurement and criteria for structure-borne sound

5.4.1   Measurement

The protection provided for a point in a studio or other quiet area from impact noise or vibration generated elsewhere cannot be measured by determining the difference between corresponding levels in the two rooms, because there is no source room as such. Instead, the accepted method, as recommended in BS2750, is to inject impacts of known energy into the structure at a point that is considered a typical point of origination of impact noise and to measure the sound level produced in the quiet area. The standard impacts are produced by dropping a series of hinged hammers, weighing 500 g each, from a height of 40 mm on to the floor or other suitable surface. It is common practice to use five such hammers, which are lifted and dropped in rotation by cams on a rotating shaft. The heads of these hammers are made of metal, but provision is made for hard-rubber caps to be fitted over the heads; these are intended to reduce the rate of rise of the contact pressure between the hammer and the floor surface.

The sound level in the receiving room is measured by means of apparatus similar to that described in Chapter 2, in octave or $\frac{1}{3}$ octave bands. The figures thus obtained may be compared with standard forms of spectrum or converted into single-figure loudness equivalents by the methods described in Chapter 2.

5.4.2   Criteria

The method of converting the octave-band figures into a single-figure loudness level yields a satisfactory criterion for most purposes. For most of the routine checking carried out by the author in the BBC studios, the band levels were summed by the method of Mintz and Tyzzer (1952), which was the most accurate method available when the criterion was being established and related to new constructions.

A tapping machine gives a succession of impacts of some 20 dB higher energy than an average footstep, with a higher relative content of high-frequency components. A loudness level of 55 phon was found to be a satisfactory criterion; if this was satisfied, real footsteps could not be heard above a background noise satisfying curve *c* of Fig. 3.1.

It is probable that a modern review of the data using the better established loudness-summation methods would give very much the same result. In general, however, more thorough methods of studying the isolation of studios against structure-borne noise are available, and these will be described in Chapter 6.

A disadvantage of the tapping machine has been pointed out by several authors, the most recent account being by Cremer and Gilg (1970). It has been mentioned above that resilient floor coverings and floating floors are equally effective against the transmission of footsteps. The tapping-machine hammers are much lighter than the limbs they are intended to represent, and in consequence give misleading comparisons between the two types of floor, the resilient covering worse than the stiff floating floor. If the mass of the hammers is increased sufficiently to give a realistic improvement for the resilient layer, the floating floor appears to be ineffective. These authors find that the best solution is to replace the hammers by an electromagnetic shaker acting through long rods. Nevertheless, all criteria for impact-noise transmission in dwelling houses, both in the UK and elsewhere, are expressed in terms of spectra generated in the receiving rooms by tapping machines in the rooms above.

Chapter 6

Measurement and analysis of sound transmission in studio centres

6.1 The problem

The methods of carrying out simple measurements of airborne-sound insulation and of the levels of transmitted impact sound produced by a standard tapping machine have been briefly described in Chapters 4 and 5. It is not sufficient, however, to regard these two forms of sound propagation as independent and capable of analysis by such simple measurements.

The transmission loss between a noisy and a sensitive area is nearly always a function of both airborne and structure-borne propagation, the relative importances of which vary widely with the situations of the two areas and the construction of the building. If the transmission loss between two such areas is found to be operationally inadequate, in spite of apparently satisfactory design and construction, measurements must be made to ascertain all the possible paths of propagation and the relative amounts of sound energy carried by them.

Consider, for instance, the two rooms represented in Fig. 4.1. Apart from the possible flanking paths that are shown, there may be communication through a common space above the ceilings or through ducts carrying ventilating air or electrical cables. The wall between the rooms may be nominally of multiple-leaf design, but the leaves could be connected or bridged by material that has been allowed to accumulate in the cavities during contruction. When measurements are made to check the performance of the partition, therefore, it is essential that they yield information on such unwanted paths of transmission or, at least, give a value for the transmission loss along a path directly through the partition.

6.2   Methods of measuring transmission loss along a single path in presence of other paths

6.2.1   Identification of paths by time and attenuation

Several methods of distinguishing the effects of alternative paths are given by Burd (1968a). The most versatile methods are those in which the paths are distinguished by their differing times of transit. Raes (1954) used very short trains of waves as the test signal and observed the transmission along paths of different delay times to the receiving-room microphone as a series of repetitions of the test impulse on an oscillographic trace. Separate paths could be distinguished, provided that the times of transit differed by more than the duration of the test signal (Fig. 6.1).

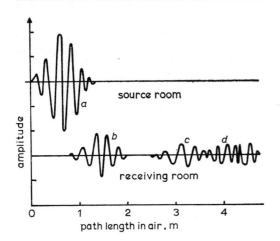

Fig. 6.1   Use of short train of waves for distinguishing between paths of propagation

a Transmitted signal
b First received pulse
c Pulse arriving by indirect path
d Subsequent pulses not resolved

The method used most extensively by Burd himself and his colleagues in the BBC depends essentially on measuring the cross-correlation function between the signals in the source and receiving rooms while a continually increasing delay is introduced into the source-room signal.

The correlation function between two time-varying signals $f_1(t)$ and $f_2(t)$ is defined as:

$$g_{21} = \frac{1}{T} \operatorname*{Lim}_{T \to \infty} \int_0^T f_1(t) f_2(t) \, dt \quad . \quad . \quad (6.1)$$

If the two functions are identical apart from a constant factor, the function g_{21} has a stationary value, equal to the average value of the product during unit time. If the two functions are similar but opposite in sign, the crosscorrelation function will be the negative of that value. Furthermore, if the signals are closely related to one another but one is delayed by a time t' with respect to the other, a high stationary value may be restored to the function by delaying the second signal by an equal amount t'.

If, now, we have equipment that gives an indication of the crosscorrelation between the signals from two microphones, one in a source room and the other in a receiving room, and if a slowly increasing delay can be applied to the signal from the source-room microphone, a maximum of the correlation function will appear for the t' corresponding to the transit time along each path of transmission. The integration over infinite time required by eqn. 6.1 is approximated closely enough if the time T of integration is made large compared with the period of the lowest sound frequency of interest; Burd used an integrating time of approximately $\frac{1}{2}$ s. Fig. 6.2 shows a circuit diagram of his apparatus. In practice, the delay is produced by recording the signal continuously on magnetic tape and reproducing with a second head

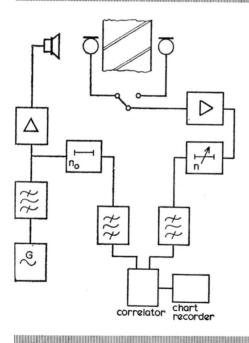

Fig. 6.2 Schematic of correlation equipment

n_0 and n in milliseconds

correlator   chart recorder

separated by a short, variable distance from the recording head. It not possible to obtain very small delays by this method, because of the finite size of the heads; a separate delay is therefore introduced into each channel, and it is the difference between the two delays that must equal the path-transit time for maximum correlation. To obtain the attenuation along each path to be obtained, the loudspeaker-drive voltage is correlated in turn with each microphone (Fig. 6.2), instead of correlating the two microphone outputs.

If wideband random noise is used as the test signal, the correlation function will fall steadily to zero as the relative delay between the two signals increases, because those components of the signal having the highest frequencies immediately fall out of phase and are followed progressively by the rest in descending order of frequency. A stage is soon reached at which statistically equal numbers of components have positive and negative correlations and the net correlation is zero.

Conversely, if a narrow-bandwidth signal is used, such as a pure tone, there will be a maximum of the correlation function at intervals of delay equal to the period of the signal. For intermediate bandwidths, the graph of the correlation function against delay becomes a damped sinewave and the output of the indicating equipment shows a series of lobes centred on each value of transit time.

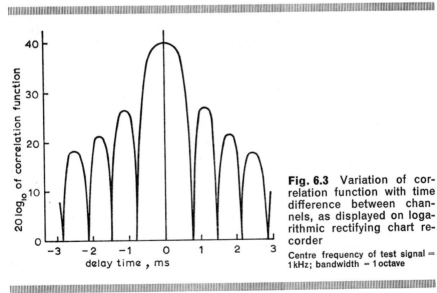

Fig. 6.3 Variation of correlation function with time difference between channels, as displayed on logarithmic rectifying chart recorder

Centre frequency of test signal = 1 kHz; bandwidth = 1 octave

A wide band of noise therefore gives precise information on transit time and good resolution between different paths but no information regarding the effect of frequency. A narrow band gives the attenuation between the microphone for an exact frequency, but has poor discrimination between times of transit and is therefore useless for identifying paths. Burd chose octave bands for general diagnostic work as a good compromise between time and frequency discrimination. Fig. 6.3 shows the calculated lobe structure for an octave bandwidth centred on 1 kHz, as displayed on a rectifying chart recorder.

The crosscorrelation method gives the same information as the short-train method of Raes; it requires more elaborate equipment, but can measure paths of higher transmission loss since more sound power can be injected in the steady signals than in short trains of waves. For the experimental details, the reader is recommended to refer to Burd (1968).

These two methods, as it will be realised, do not identify the paths of transmission, but only their transit times and transmission losses. It is necessary to infer the actual paths from these data in conjunction with a knowledge of the structure of the building.

6.2.2 Spatial identification of transmission paths

The tracing of paths distinguished by their transit times and transmission losses may be assisted by several methods. A small loudspeaker fitted with a protective hood may be scanned over the surface of a partition in an attempt to find weak places in the sound insulation. This method is often recommended, but, in the author's experience, the standing-wave patterns in a typical studio generally mask the variations in sound level that could indicate such weak spots.

The presence of severe flanking transmission through the side walls may sometimes be detected from the difference between the mean sound-pressure level along the partition wall in the receiving room and that in the body of the room, away from the wall. In the absence of flanking transmission, this amounts to about 3 dB but the difference diminishes in the presence of radiation from the side walls. Severe flanking transmission is sometimes revealed by measurements of the sound-transmission loss between the source room and a room beyond the receiving room. Without flanking paths, this is likely to approach the sum of the measured transmission losses in the two intervening walls, but, when flanking transmission is present, the decrease in the level of transmitted sound in the further room is comparatively small, representing only the attenuation along the additional length of flanking wall (Fig. 5.3).

6.2.3   Use of accelerometers in sound-transmission analysis

Undoubtedly, the most powerful tool in the investigation of sound-transmission paths in a building is the accelerometer. This was mentioned by Meyer *et al.* (1951). Accelerometer methods have since been developed for many types of investigation by Ward (1962). The most suitable type of accelerometer consists of a small metal disc connected to a baseplate by a disc of a piezo-electric material (Fig. 6.4). The baseplate is fastened to surfaces in the building, the most convenient attachment being a thin layer of modelling clay or beeswax.

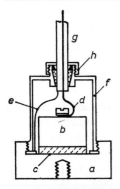

Fig. 6.4   Piezoelectric accelerometer

   a Base plate
   b Steel seismic mass
   c Piezoelectric material
   d Signal lead
   e Earth lead
   f Screening can
   g Coaxial cable

The force on the piezoelectric material, and hence the voltage across it, is proportional to the mass of the metal disc and the acceleration imposed on it. The voltage is applied to an amplifier with a suitably high input impedance and low output impedance; the output is then treated as a normal microphone signal and applied to a measurement chain (Fig. 2.1). The preamplifier may include an integrating stage to convert the output to a voltage proportional to the vibration velocity instead of the acceleration amplitude.

Accelerometers may be used for three main purposes:

(*a*)   the identification of sources of noise in a building, and of the principal paths of transmission of the noise through the structure

(*b*)   the prediction of sound-pressure levels in enclosures from the acceleration amplitude of the structure, and hence

(c) the measurement of the sound reduction of partitions in the presence of flanking paths.

6.2.3.1 Identification of sources

For the identification of sources and transmission paths, it is usually more useful to have measurement techniques that quickly produce an indication of the nature of the problem and of the orders of magnitude than to have techniques that, while being more accurate, involve more labour and time.

For diagnostic work in a building, the test signal may be a vibration of variable frequency or narrow-band random noise. Alternatively, it may consist of impacts produced by means of a standard tapping machine or a single blow from a mass allowed to fall through a known distance on to a hard surface. Accelerometers are then used to measure the amplitudes of vibration at chosen points in the building.

It is usual also to mount an accelerometer close to the generator of the test sound to provide a control signal.

For the identification of existing noise sources, one may compare sound levels in different parts of the structure, playing 'hot and cold', so to speak, or compare the character of the interfering noise, or the shape of its spectrum, with that given by an accelerometer placed near various possible sources in the building. This method is particularly useful, because the sound transmitted by the source into the structure usually differs in character from that radiated into the air, thereby making direct aural recognition difficult.

6.2.3.2 Paths of transmission

The principal paths of transmission may be identified by similar methods, so that remedial measures may be prescribed. Such measures may consist of the application of antivibration mountings to machinery, resilient joints in the structure, or soft floor coverings.

As an example, Ward quotes a basement studio which was subject to footsteps from a pavement adjacent to the building. Comparison was made between the acceleration levels at the pavement on the one hand and on a wall common to the faulty studio and adjacent studio on the other. The footsteps were virtually inaudible in this latter studio. Measurements were made also in positions between the pavement and the two studios. The results (Fig. 6.5) revealed that the difference in the transmission of sound from the pavement to the two studios occurred entirely across a wall separating the two studios from the corridor; it was concluded from this that the part of the wall bordering the unaffected studio was of cavity construction, whereas that part along the noisy studio was solid. An extra leaf was therefore added to the wall of the affected studio along this section, and this greatly reduced the noise in the studio. The leaf was made of soft fibreboard and plasterboard nailed together on timber studding. Residual radiation of the sound from the studio ceiling, which was supported by the solid part of the wall, was diminished to a satisfactory level by damping the void with rockwool.

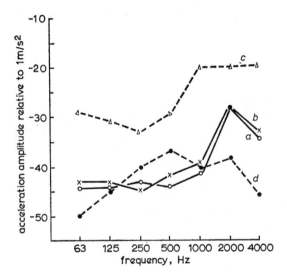

Fig. 6.5 Accelerometer measurements of vibration on two sides of a wall in path of propagation (Ward, 1962)

a Source side at point where wall has solid structure
b Receiving side at same point
c Source side where wall is discontinuous
d Receiving side at same point as *c*

6.2.3.3 Estimation of sound-power levels from acceleration levels in studio boundaries

There are many instances in which the sound-pressure level of the sound in the studio from ground vibrations due to traffic on road or rail must be estimated before any part of the building or its foundations are constructed. Alternatively, as in the construction of the extension to the BBC Broadcasting House, the basic structural shell may be partly complete, but the measures for the isolation of the studios from the underground train or other noise must be designed in advance of the completion of the main walls separating one studio from another.

The estimation of such sound-pressure levels may be made with satisfactory accuracy by assuming that the radiation efficiency of the surfaces is unity at frequencies above the critical frequency of the bending waves in the wall material. With brick and concrete structures, this frequency is low in the audio range, so that the pressure levels may be predicted accurately over most of the desired range, and approximately even at frequencies below about 200 Hz.

If the vibration velocity of a surface is represented by

$$v = v_0 \sin\omega t \qquad . \quad . \quad . \quad . \quad . \quad (6.2)$$

where v_0 is the velocity amplitude, the acceleration is given by

$$\alpha = \omega v_0 \cos\omega t \qquad . \quad . \quad . \quad . \quad . \quad (6.3)$$

The sound-pressure near the surface is given by

$$p_0 = \rho c v_0 = \rho c \alpha_0 / \omega \qquad . \quad . \quad . \quad . \quad (6.4)$$

where α_0 is the acceleration amplitude.

Above the critical frequency f_c and in the absence of damping, the efficiency of radiation of an infinitely large surface is given by Gösele (1953) as

$$s = \{1 - (f_c/f)\}^{-\frac{1}{2}} \qquad . \quad . \quad . \quad . \quad (6.5)$$

where f is the frequency. Below f_c, the radiation is zero.

For finite, damped walls, the efficiency lies between these values and unity; by assuming unity efficiency from f_c upwards, the near sound field may be predicted with a very satisfactory degree of accuracy.

6.2.3.4 Measurements of transmission loss of partitions

Accelerometer measurements may, for example, be used for measuring the sound-transmission loss of a partition in the presence of flanking sound. The sound-pressure level on the source side of the partition is measured in the conventional manner, and the average r.m.s acceleration of the surface of the panel is measured by means of accelerometers fixed to the partition on the receiving-room side.

The near-field sound-pressure level is then derived from the accelerometer measurements, and is used in place of the sound-pressure level measured directly by means of microphones.

If we assume that the microphone in Fig. 2.1 is replaced by an accelerometer and preamplifier that together give an output of k millivolts for g peak acceleration, the equivalent s.p.l. for an r.m.s. output-voltage level V (relative to $1\,V$) is given by

$$L = V + 207 - 20 \log_{10} kf - G \quad \cdot \quad \cdot \quad \cdot \quad (6.6)$$

where f is the frequency and G the net voltage gain between the preamplifier output and the indicating meter.

The advantage of this method over the conventional method is that the accelerometers indicate only the actual movements of the panel as a result of the sound pressures on the other side, and it is clear that, to a first order, this indication is unaffected by sound energy that arrives from the source room by parallel flanking paths through side walls or cracks in the wall surrounding the measured partition.

Fig. 6.6 illustrates Ward's measurements on a steel door built into an aperture by normal building techniques. Curve a shows the transmission loss measured by the conventional method, and curve b shows the transmission loss measured in the same way after sealing the edges of the door with soft modelling clay. It will be seen that curve a is considerably lower than curve b because of the transmission of sound through small cracks round the junction of the door with the aperture. Curve c was obtained from accelerometer readings instead of microphone measurements in the receiving room, and it is clearly unaffected by the leakage.

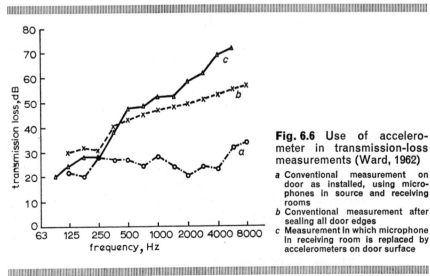

Fig. 6.6 Use of accelerometer in transmission-loss measurements (Ward, 1962)

a Conventional measurement on door as installed, using microphones in source and receiving rooms

b Conventional measurement after sealing all door edges

c Measurement in which microphone in receiving room is replaced by accelerometers on door surface

6.3 Experimental work using scale models

A considerable amount of work has been expended on the perfection of scale models for the representation of auditoria and their acoustic properties, a ratio of between 1:10 and 1:8 being generally favoured; further reference to this will be made in Chapter 10. However, very little attention seems to have been paid to the possibility of scale-model techniques for the routine evaluation of designs with regard to sound insulation and transmission through large structures.

The potential advantages of such scaling techniques are very great, even for such relatively large linear scales as one-quarter. The simplest and most obvious scaling scheme is that proposed by Kosten (1949), in which time is scaled inversely as the linear dimensions, while densities of, and velocities in, the materials remain unaltered. The models can be made of the original materials since all the essential properties scale correctly. Meyer (1956) used models for verifying predictions of the attenuations produced by discontinuities in solid conducting paths. Kuhl and Kath (1963) tested in scaled form sound-insulating wall cladding and other components for the studio centre at Hanover in Germany, but the possibilities of the technique as a means of predicting the transmission losses from one point in a building to another does not seem to have been exploited to the extent one might expect. It could, if proved reliable, also be employed to predict the effect of remedial measures applied to a building without the necessity of carrying out experiments in full size.

There are several possible difficulties. One is that, although nearly all the main properties of a building can be scaled without any error, misleading results could appear as a consequence of anomalous behaviour of granular or fibrous building materials when working at scaled frequencies. The internal damping of such materials, for instance, may vary with frequency. Kuhl and Kaiser (1952) found that building elements filled with dry sand show a marked change of damping factor with amplitude of excitation. Thus application of scale models to the study of sound transmission must be founded on a complete knowledge of such anomalies.

Another difficulty, pointed out by Kosten (1949), is that external constraints on a building, such as those due to the force of gravity, may not necessarily be satisfactorily scaled and yet may contribute to the sound-transmission properties of joints and large loaded spans.

Thirdly, the fabrication of sufficiently accurate scale models of building components presents considerable difficulty in handling and processing materials.

Two university departments of building are, however, working on these problems. Recent work at Aston by the author and a colleague (Gilford and Gibbs, 1971) has shown that in metals and in building materials generally, the loss factor changes only slowly with frequency. The loss factors were determined either by means of mechanical admittance measurements or from rates of decay of vibration; they varied from 10^{-4} in steel to rather over 10^{-2} for foamed concrete and wood fibreboards. Loss factors of this order are of some importance in determining the transmission losses of partitions, but have less influence on the propagation of sound through building structures. Small changes in frequency are therefore not likely to affect the accuracy of the agreement between the model and fullsized versions of a structure.

Mackenzie (1970), working at Heriot-Watt University in Edinburgh, has shown that some materials that are difficult to fabricate on a small scale may be replaced by more convenient materials, e.g. plasterboard by methyl methacrylate.

Chapter 7

Reverberation

7.1 Introduction and definitions

Chapters 7–9 are concerned with the behaviour of sound energy that is temporarily confined within a studio or other enclosure, how its confinement within the space affects its character and the way in which its energy is dissipated.

No apology is required for introducing the idea and definition of reverberation time at the outset, since this concept is still the most important guide to acoustic design. Historically, it was also the first feature of room acoustics to be defined and measured. Sabine (1922) originally defined it as the time taken for sound in

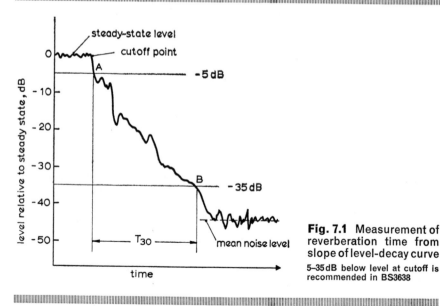

Fig. 7.1 Measurement of reverberation time from slope of level-decay curve

5–35 dB below level at cutoff is recommended in BS3638

an enclosure to decay to inaudibility from an original intensity one million times greater. It is now generally defined as the time occupied by a decrease of 60 dB in the sound-pressure level, but is normally measured over a smaller change of level. Indeed, the relative British Standard (BS3638) specifies that it should be measured over a range of only 30 dB, starting at a level 5 dB lower than the initial steady state. This recommendation is represented in Fig. 7.1. If T_{30} is the time taken for the level of fall from point A to point B, the reverberation time is defined as $2T_{30}$.

These methods are based on the assumption that the energy of the sound in the enclosure decreases at a rate proportional to the instantaneous energy density. In practice, this is broadly true, though the sound pressure may fluctuate widely from the exponential time relationship that would result from an exact proportionality. Energy is lost from an enclosed sound field, mainly by absorption from the boundary surfaces, and, since any wavefront will travel from one surface of the enclosure to another between successive losses of energy, it is clear that the decay of energy must be a discontinuous and irregular process, statistical in its general nature and characterised by fluctuations of varying amplitude. No tidy analytic solution of the decay process is therefore to be expected.

There are three main methods of analysis which have had varying degrees of success, and which may be applied alternatively to particular problems:

(*a*) statistical theory
(*b*) geometrical theory
(*c*) wave theory.

7.2 Statistical theory of reverberation: reverberation in large spaces

The sound field in the room is regarded as consisting of a very large number of wavefronts travelling in different directions and being scattered and partly absorbed by numerous reflections at the boundaries until they eventually become dissipated. If there is a constant source of sound present, the intensity in the enclosure will increase steadily until it reaches a value at which the rate of absorption of energy at the boundaries is equal to the rate of production at the source.

E

7.2.1 Calculation of reverberation time by statistical methods

Consider the mass of wavelets travelling in all directions, as described above, and being sustained by a source of energy of power P. Then, if I is the intensity of sound energy in the enclosure and V is the volume of the enclosure, the equation for the growth of intensity is

$$V\frac{dI}{dt} = P - P_a \quad . \quad . \quad . \quad . \quad . \quad (7.1)$$

where P_a is the rate of absorption of energy by the boundaries. Some energy is also dissipated in the passage of the sound through the air as well, but this will be ignored for the present. With this reservation, we may consider P_a as made up from the absorption of energy from wavefronts of infinitisemal area δS, a large number of which strike the boundaries during any small element of time. If one of the wave elements has a thickness δl perpendicular to the wavefront, its volume will be $\delta S \delta l$, and it will contain an amount of energy $I\delta S \delta l$. If a fraction α of this energy is absorbed at a reflection at the boundary, the boundary is said to have an **absorption coefficient** α. If $\bar{\alpha}$ is the mean absorption coefficient of the boundaries of the enclosure, the element will lose an amount of energy, at each reflection, given by $I\bar{\alpha}\delta S \delta l$, on the average, every time it is reflected, and this will occur at intervals of l_m/c, where l_m is the mean free path in the enclosure.

Kosten (1960) has shown that the mean free path in an enclosure of any shape is given by

$$l_m = 4V/S \quad . \quad . \quad . \quad . \quad . \quad (7.2)$$

where S is the total surface area of the boundaries of the enclosure.

The energy lost by the element in unit time is therefore

$$I\delta S \delta l \bar{\alpha}(cS/4V)$$

and therefore the loss of energy through the volume is

$$P_a = IcS\bar{\alpha}/4$$

Inserting this value into eqn. 7.1, we obtain

$$V(dI/dt) + \tfrac{1}{4}cS\bar{\alpha}I = P$$

which yields

$$I = 4/(cS\bar{\alpha})P\{1 - \exp(-\tfrac{1}{4}VcS\bar{\alpha}t)\} \quad . \quad . \quad . \quad (7.3)$$

This tends asymptotically to a steady intensity, given by

$$I_\infty = 4P/(cS\bar{\alpha}) \quad . \quad . \quad . \quad . \quad . \quad (7.4)$$

If, when the steady state has been reached, the source is cut off so that P becomes zero, the equation of the decay becomes

$$VdI/dt = -IcS\bar{\alpha}/4$$

and therefore

$$V \log_e(I/I_0) = -cS\bar{\alpha}t/4 \quad . \quad . \quad . \quad . \quad (7.5)$$

The reverberation time, defined as the time taken for I/I_0 to fall to 10^{-6}, is given by

$$T = 4V/(cS\bar{\alpha}) \log_e 10^6 = 13\cdot8 \times 4V/cS\bar{\alpha} \quad . \quad . \quad (7.6)$$

Putting $c = 340\,\text{m/s}$, we finally obtain

$$T = 0\cdot162V/S\bar{\alpha} \quad . \quad . \quad . \quad . \quad (7.7)$$

(where T is seconds, V in cubic metres and S in square metres).

This is known as Sabine's formula, after Sabine (1922).

Notice, from eqn. 7.5, that the logarithm of the intensity diminishes linearly with time; this is the basis of nearly all practical methods of measuring reverberation time, as was implied at the start of this chapter.

Fig. 7.2 shows the form of the statistical increase of intensity after switching on a source of constant power and the subsequent decay of intensity when the source is switched off. In Fig. 7.2a, these curves are shown on a linear scale, and, in Fig. 7.2b, on a logarithmic scale of intensity.

Sabine's simple equation serves as an accurate method of calculating the reverberation time of an enclosure in which truly statistical conditions can be maintained during the period of the decay. This is so if $\bar{\alpha}$ is small in comparison with unity so that the wave front undergoes many reflections before virtual extinction. If, however, we imagine an enclosure with perfectly absorbing boundaries so that $\bar{\alpha} = 1$, the equation indicates a reverberation time of $55\cdot2V/cS$. Since $4V/S$ is the mean free path of the enclosure, this means that the sound wavefront that we are following will suffer 13 reflections during the time of reverberation, which is impossible, since we are assuming perfect absorption. The explanation of this paradox is that we have assumed, in effect, that the starting condition is one with a large number of wavefronts all travelling independently outwards and striking the boundaries. Their subsequent history, from the point of view of the enclosure, is no different from that which would have been had the perfectly absorbing wall not been present. The attainment of an intensity that is 10^{-6} of the starting intensity would therefore have taken place after a time that is independent of the actual dimensions of the enclosure.

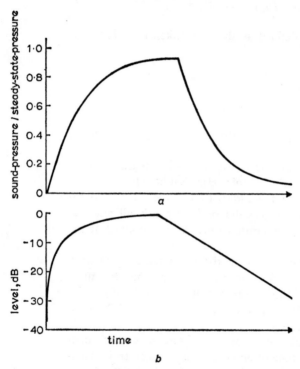

Fig. 7.2 Ideal forms of rise and decay of sound in an enclosure
a Linear scale of intensity
b Logarithmic scale (intensity level)

These considerations prompted Eyring (1930) to study the implications of the assumptions made above, and, in particular, that of the continuous dissipation of energy. Let us follow the course of a single element of wavefront, with initial intensity I_0, as it traverses the enclosure. It is reflected from the first boundary that it meets, and its intensity is reduced to $I_0(1-\bar{\alpha})$. After a subsequent interval of average length $4V/Sc$ it suffers a further loss of energy, so that its intensity is now $I_0(1-\bar{\alpha})^2$, and, after k such reflections, occupying a total time of $4kV/cS$, its intensity will have fallen to $I_0(1-\bar{\alpha})^k$.

We may therefore write

$$I = I_0(1-\bar{\alpha}) \frac{cSt}{4V} \quad \cdots \quad \cdots \quad (7.8)$$

Taking logarithms of both sides,

$$\log_e(I/I_a) = (cSt/4V) \log_e(1-\bar{\alpha})$$

and, if $\log_e(I/I_0) = -13\cdot 8$, as in the derivation of Sabine's formula, we finally have

$$T = 4V/\{cS \log_e(1-\bar{\alpha})\}$$
$$= 0\cdot 162V/\{-S \log_e(1-\bar{\alpha})\} \quad \cdots \quad (7.9)$$

(in SI units).

This is known as Eyring's equation; it differs from Sabine's only by the substitution of $-\log_e(1-\bar{\alpha})$ for $\bar{\alpha}$.

Now

$$-\log_e(1-\bar{\alpha}) = \bar{\alpha}+(\alpha^3/3)+(\alpha^5/5) \quad \cdots \quad (7.10)$$

and hence, for small values of $\bar{\alpha}$, the equation becomes equal to Sabine's equation. For large values of $\bar{\alpha}$, the value of $-\log_e(1-\bar{\alpha})$ becomes infinite, so that the calculated reverberation time is zero. Eyring's formula should, therefore, be used for all enclosures with a high average absorption coefficient; it is found to give more accurate results in most circumstances. It is, of course, more convenient and equally viable to use Sabine's formula for lightly damped enclosures, a convenient limiting value of the mean coefficient being $0\cdot 25$. For this value, the difference between the two formulas amounts to approximately 2%, and an error in the designed reverberation time of this amount is not subjectively significant.

In applying Sabine's formula, it is simply necessary to derive the denominator by adding together all the separate contributions of absorption obtained by multiplying the areas of the various types of surface by their absorption coefficients. To this sum is added an amount representing absorption by the air as the sound traverses it between reflections. This is given by $4mV$, where V is the volume of the studio and m is the fraction of energy lost by a plane wave in unit distance. The most accurate available values of m are those of Evans and Bazley (1956).

The effective absorption of people, furniture etc. must be added. All these sources of absorption will be considered in Chapter 8.

In applying Eyring's formula, the mean absoption coefficient $\bar{\alpha}$ is obtained by dividing the total absorption (sum of products of areas and absorption coefficients) by the total surface area, and the function $-\log_e(1-\bar{\alpha})$ is obtained from tables and multiplied by the total area (Table 13.3). The absorption contributions of the air, people, furniture etc. are then added to give the denominator.

In the derivations of both Eyring's and Sabine's formulas, it is assumed that the sound field in the studio is perfectly diffuse, i.e. the sound intensity is uniform over the whole volume of the room and particle velocities are randomly distributed in direction at any moment. This requires, in turn, that the absorbing surfaces should be scattered about the room without any systematic order. These requirements will be examined closely in Chapter 9. If the sound field is not diffuse, the decay of sound will not be according to an exponential law, and the reverberation time cannot be calculated by either of the formulas given above. Such cases lie in the province of geometrical or wave theory, and they will be considered briefly in Sections 7.3 and 7.4.

7.2.2 Subjective consequences of reverberation

The reverberation in a studio has several important effects on the programme output, most of which are dealt with in textbooks on auditorium or studio acoustics. Briefly, excess of reverberant energy in the sound field causes confusion, lack of definition or clarity in music, poor intelligibility of speech, and perhaps **colouration,** the accentuation of particular frequencies in speech or music so that certain notes or vowel sounds assume unnatural prominence.

Deficiency in reverberation causes inadequate loudness at a distance from the source, poor or harsh tonal quality, a lack of rhythmic urge in music and an impression of scrappy playing by instrumentalists. Excess or deficiency in particular frequency regions can be heard as a corresponding unbalance of the tonal structure.

It was realised early in the history of acoustics that there was an optimum value or range of values of the reverberation time. Many recommendations have been made from time to time, normally linking reverberation time with studio volume and programme type.

Three methods have normally been used for establishing these optima. Early attempts were made by various authors to establish optimum reverberation/frequency characteristics and mean reverberation times from arguments based on the theories of hearing. It has been argued, for instance, that the frequency variation of reverberation time should be such that all the components of a wideband sound, all starting exactly 60 dB above the threshold of audibility, fall to the threshold of hearing at the same instant. This assumption, together with some other apparently relevant data, led to the publication of an optimum reverberation/frequency characteristic shaped something like an equal-loudness contour, quite regardless of the agreed opinion that the sound of a studio with a characteristic remotely resembling this is quite unacceptable. Unfortunately, these recommendations received very wide acceptance for many years, and are still occasionally quoted. There have been other attempts, based on equally untenable assumptions, to derive optimum reverberation times and frequency characteristics from supposed first principles, but no good purpose would be served by describing them in detail.

A number of serious studies have been made by presenting recorded extracts of programmes with various conditions of reverberation to teams of sensitive and experienced listeners, noting their comments and hence reaching a consensus from which the optimum design parameters can be derived. Kuhl (1954) used extracts of music from various historical periods recorded in studios and concert halls with various reverberation times, and he submitted the recordings to a panel of listeners. He found that, irrespectively of the volume of the enclosure, a reverberation time of 1·5s was most suitable for 18th-century music and for modern composers, while 2·2s was ideal for music of the 'romantic' period, in which larger orchestras and denser orchestration were usual. He proposed 1·7s as the best compromise for music of all kinds. In any such definitive experiment, however carefully done, it is generally possible to raise doubts about the validity of the results. One could say, in this instance, that the results depended too much on the placing of the single omnidirectional microphone used for the recordings. Moreover, the available studios and concert halls nearly all had reverberation times near the top or bottom of the range investigated, there being very few with times of 1·6–2s. Any direct subjective experiment inevitably involves compromises or contradictions, and, although Kuhl carried out this series with considerable care and excellent equipment, his compromise figure of 1·7s is now generally regarded as being too low, particularly for large studios or auditoria.

The third approach is the accumulation of opinion over a long period of time from a variety of sources and correlation with measured reverberation times for all the studios used for the subjective data. The subjects should have preferably taken part in the productions of broadcasts or recordings from some of the studios. Surveys of this kind have been carried out by Somerville (1953) and Somerville and Head (1957) with concert halls and broadcasting studios, by Beranek (1962, pp. 425–431) for concert halls and opera houses and by Burd, Gilford and Spring (1966) for sound and television studios. The results will be further considered in Chapters 11 and 12, which deal with the design of sound radio and television studios. To summarise, one may say that, in general, for a 'natural' or distant single-microphone balance,

(*a*) the optimum reverberation time increases steadily with the volume of the studio, the best average time for different types of music reaching a maximum of about 2s for the largest concert hall or studio

(*b*) the optimum time is greatest for organ music and least for speech; other forms of programme have intermediate requirements.

7.2.3   Liveness

The **liveness** at a point in a sound field is the ratio of the intensity of the reverberant-sound field at the point to that of the sound coming directly from the source (Maxfield and Albersheim, 1947).

It was shown in Section 2.3.4 that this ratio may be written as

$$16\pi\, r^2(1-\bar{\alpha})/S\bar{\alpha}$$

where $\bar{\alpha}$ is the mean absorption coefficient of the surfaces and r is the distance of the point from the centre of the source of sound.

In a studio of about optimum reverberation time, good broadcasts are obtained with liveness ratios between about 5 and 8. It has been found possible, on occasions, to make comparatively satisfactory recordings from a strange concert hall, without any opportunity for rehearsal, by suspending a single microphone at a point calculated, on the basis of reverberation measurements, to have a liveness ratio within this range.

7.3   Reverberation in small studios: room modes

The statistical view of reverberation is, as stated above, only valid for studios with dimensions which are large compared with the wavelengths of the programme sounds and in which the sound field is diffuse. It is necessary, therefore, to deal with small studios, such as those generally used for speech programmes, in a different manner. The sound fields in small rectangular rooms have been exhaustively examined by many authors. Rayleigh (1896) showed that the air mass in a rectangular room has an infinite number of natural models of vibration with frequencies given by the well known equation

$$f = \frac{2\pi}{c} \left(\frac{n_1{}^2}{l_1{}^2} + \frac{n_2{}^2}{l_2{}^2} + \frac{n_3{}^2}{l_3{}^2} \right)^{\frac{1}{2}} \qquad . \quad . \quad . \quad (7.11)$$

in which l_1, l_2 and l_3 are the lengths of the sides of the room and n_1, n_2 and n_3 are zero or positive integers.

The number of frequencies represented by this expression increases very rapidly with the upper limiting frequency of the sound. Bolt (1938–39) pointed out that each modal frequency may be represented by a point with co-ordinates (n_1/l_1, n_2/l_2, n_3/l_3) in 3-dimensional space. These points form a rectangular lattice in what he calls frequency space, and the number of modes between any two frequency limits is the number of points lying in an octant of a spherical shell bounded by spheres centred on the origin of co-ordinates of radii corresponding to the two frequency limits. The total number of modes below a particular frequency therefore increases as the cube of that frequency, and the number within a specified bandwidth increases as the square of the centre frequency. Even a small talks studio may have several modes within a bandwidth of 1Hz at, say, 300Hz.

However, it is a matter of common observation that only a few frequencies at most become prominent enough to be heard as colourations on speech from a small studio. The reason for this is partly, but not yet fully, understood.

In physical terms, the Rayleigh expression embraces three types of mode:

(*a*) those in which two of the *n*s are zero. In this case, the modal frequencies form an arithmetic series given by

$$f_m = cm/2\pi l_n \qquad . \quad . \quad . \quad . \quad . \quad (7.12)$$

where *m* is any positive integer.

These modes are characterised by particle velocities that are always parallel to one set of room edges, with antinodes of pressure over one pair of opposite walls. They are known as **axial modes**

E*

(*b*) **tandential** modes, in which one *n* is zero and the particle velocities are parallel to a pair of opposite surfaces

(*c*) **oblique** modes, in which no *n* is zero and the particle velocities are oblique to all the wall surfaces and edges.

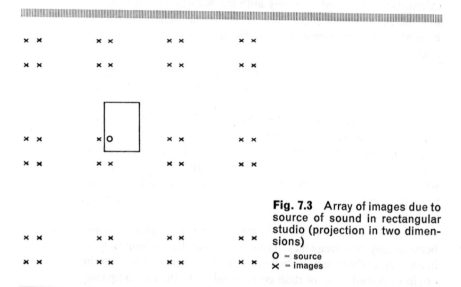

Fig. 7.3 Array of images due to source of sound in rectangular studio (projection in two dimensions)

O = source
✕ = images

A careful comparison of the characteristics of these types was made by Mayo (1952). He calculated the rates of buildup and decay of the reverberant-sound pressures from individual images and groups of images of the source, formed by reflection at the walls (Fig. 7.3). He showed that the early reflections at a point in the studio were randomly phased, since they came from individual images of which the distances from the point were not regularly spaced, but that, after a short interval, regular series of prominent frequencies started to appear from rows or planes of regularly spaced images. The 'line' frequencies, i.e. those formed by lines of images, build up rapidly, but decay rapidly, owing to attenuation with distance and absorption at repeated reflections. The 'plane' frequencies, which are those given by the Rayleigh formula, are formed more slowly, but also decay very much more slowly, because, being formed by plane arrays of images, there is no attenuation with distance other than that due to surface absorption. Mayo showed that the only frequencies that are likely to be subjectively significant are those representing both plane and line arrays of images. These are the axial frequencies. The author

(Gilford, 1959) examined the consequences of this finding in relation to the audibility of colourations on speech in small studios. His conclusions were that, to be audible as a colouration on speech, an axial mode must

(a) be separated from axial modes on either side of it by approximately 20 Hz or more (a group or pair of modes very close to each other in frequency would behave as a single mode for this purpose)

(b) coincide with a fundamental or first formant of at least one vowel sound and be in the region of greatest speech-energy output.

Fig. 7.4 shows the frequency distribution of subjectively determined colourations in small studios; it will be seen that they occur mostly at 75–200 Hz, with a subsidiary peak at 250–300 Hz. Peterson and Barney (1952) found that the fundamentals for male speech are spread around 130 Hz, and that the first formants start at 270 Hz. These figures are consistent with Fig. 7.4. Colourations below 80 Hz are rare because of the small speech-energy density, and they disappear above 300 Hz because of the great increase in the energy contained in oblique and tangential modes at high frequencies (Table 7.1).

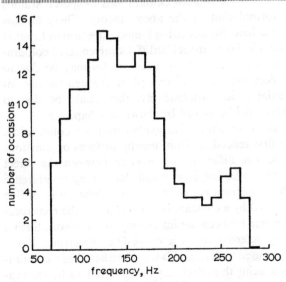

Fig. 7.4 Distribution of 61 observed colourations in small studios

Table 7.1   Numbers of axial, tangential and oblique modes of a typical talks studio

Frequency limits	Axial	Tangential	Oblique	Total
Hz				
0–50	2	0	0	2
50–1000	3	6	1	10
100–150	4	11	7	22
150–200	3	19	16	38
200–250	3	19	29	51
250–300	4	30	51	85
300–350	2	34	69	105
350–400	3	46	98	147

Individual frequencies of colourations within the population represented in the data of Fig. 7.4 could be identified, in the great majority of cases, with axial modes or groups of modes that were widely separated from their nearest neighbours.

The correlation between the colouration frequencies and the widely separated modes cannot be claimed to be perfect, though the systematic avoidance of such isolated axial modes in the voice-fundamental and first-formant regions by careful attention to room dimensions appears to have resulted in a considerable reduction in the incidence of such colourations in speech studios built since the formulation of the above theory. These studios include those in the BBC Broadcasting House extension in London and in the BBC provincial centre at Cardiff. Occurrences of colourations not explained by reference to axial modes may be due to chance coincidences between line and plane frequencies not included in the axial series. Alternatively, they could be due to insufficiently diffuse fields, as will be shown in Chapter 9.

Kuhl (1965) gives an alternative explanation for colouration, attributing it to first reflections from nearby surfaces or junctions between surfaces. The difference of pressure between the direct sound and the first reflection is so small that strong interference will result, with a spectrum consisting of an infinite series of line frequencies separated by a constant interval equal to the reciprocal of the path difference between the interfering sound waves. Such a spectrum, usually referred to as a comb-filter spectrum, has a distinct pitch associated with it, and is therefore heard as a colouration. Surfaces producing this effect may be untreated walls, observation windows or the angles between walls and ceiling, which act as

2-dimensional corner reflectors and reflect sound back in the opposite direction to that from which it came. By systematically treating suspected reflecting surfaces with patches of sound-absorbing material or with curved scattering surfaces, Kuhl reduced very substantially the colourations in a large number of small studios, without greatly reducing their reverberation times. He also quotes, in support, early experiments by Venzke (1954), who found that a microphone placed near to a corner of a studio treated with a sound-absorbing material gave more reverberant and coloured speech quality than when placed in a diametrically opposite corner not so treated. The reverberation time of the studio was, of course, common to both positions. A related effect has been reported by Gilford and Harwood (1969), who showed that colourations affecting the monitoring loudspeakers in television control rooms could be cured by treatment of the neighbouring surfaces, but they also found that the colourations were confined to positions very close to the corners of the rooms.

The possibility of colourations having their origin in early reflections must therefore be borne in mind when undertaking remedial treatment, but, in the author's experience, the axial-mode explanation appears to hold in the majority of cases. The pitch of a colouration formed by an early reflection would be expected to rise continuously as the microphone or speaker approached the nearest surface, but one tends to hear the same pitch from many positions in the studio, and the ear hears the colourations as a characteristic of the studio as a whole, rather than of a particular position.

Whichever explanation is accepted, it is true to say that the reverberation in a small room is heard not as a prolongation of individual sounds, as in a large hall, but as frequency distortions that impair the quality of speech if not eliminated by careful design or treatment.

Before leaving the subject of colourations, it should be noted that colourations can also be caused in studios of any size by the mechanical resonance of a variety of objects, from fire extinguishers to undamped ceiling structures. Such resonances can have very long decay times, and Jones (1967c) has shown that, when this is the case, they are subjectively noticeable if excited to a level 28 dB or less below the room excitation level in the presence of speech.

Care must be taken to avoid all resonant structures or studio equipment that have decay times appreciably longer than the reverberation time of the studio.

7.4 Geometrical analysis of sound in a room

In Section 7.3, we examined the possibility of frequency distortions in small rooms caused by the interference of the direct sound with wave trains reflected from relatively large surfaces.

In the same manner, the decay curve of sound in a large studio may deviate from a steady exponential course by reason of the effects of large absorbing or reflecting areas which distort the temporal pattern of the decay. For simplicity, let us regard the decay curve as being always displayed on a logarithmic scale of pressure (i.e. as sound-pressure level). The exponential time–pressure law then becomes a straight line, and deviations from the ideal law become curvatures in the line or fluctuations from it.

These effects may be studied by the methods of **geometric acoustics**, in which the propagation of sound from one place to another is regarded as occurring in straight lines analogous to the rays of light in geometrical optics. At a reflection, the incident and reflected rays lie in the same plane as the normal, and make equal angles with it. Where the rays intersect, the intensity of the sound field is equal to the sum of the intensities of the individual rays; effects depending on diffraction or interference are disregarded. This presupposes that the dimensions of the enclosure are large in comparison with the wavelength of the lowest-frequency sound with which we are concerned, and the method is therefore restricted in its application to sound of high enough frequency to fulfil this condition in a given enclosure.

The principal application of this method is in modifying the propagation of sound in a studio or auditorium to obtain certain desired effects. For instance, the shape of an auditorium may be designed to increase the sound level at points distant from a source on the stage by surrounding the stage with large reflecting surfaces so placed as to reflect a high proportion of the sound towards the rear of the hall as a parallel or convergent beam. Alternatively, as in many concert halls built since the Second World War, the first-reflected sound may be directed towards the audience to increase the contribution of the early sound in comparison with that of the reverberant sound in all parts of the auditorium. As shown in Chapter 1, the ear and brain combine to integrate all reflections arriving less than 50 ms after the first sound and to accept all these early reflections, at any rate in the horizontal plane, as coming from the same direction as the original source. Fig. 7.5 shows an example of this principle as applied to the design of the Salle Pleyel in Paris (Andrade, 1932). Both the plan and vertical-section forms were designed to increase the loudness of the sound at the back of the hall by concentrating early reflections in this area, and, by hindsight, we realise that it is the reflections within 50 ms which have this effect.

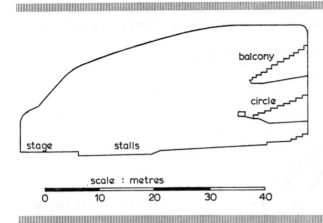

scale : metres

0	10	20	30	40

Fig. 7.5 Long section of Salle Pleyel, Paris

Unfortunately, this concentration of the first reflections of the direct sound has unwanted side effects. In a concert hall, the audience is the only efficient absorbing material present in large areas, and thus a large proportion of the sound energy from the source is absorbed during a time interval that is very short compared with the reverberation time of the hall. Fig. 7.6 shows the decay curve *a* of such a hall compared with that in an enclosure without any such direction of the first reflections. Although the reverberation time of the hall, as measured from the mean slope of the decay curve *a*, is not very different from that of the uniform slope *b*, the relative level of the reverberant sound is much lower in the first case, and the hall possesses the characteristic of 'dryness', because the reverberation is swamped by the sound coming from the direction of the source, either directly or after one reflection.

For example, if music of a continuous nature is being performed, the listener hears reverberant sound only in rests or pauses or at the end of a movement. Another disadvantage is that the Haas effect does not cause the direct sound to dominate the impression of direction in the vertical plane as it does in the horizontal plane (Somerville, *et al.*, 1966) with the result that the apparent source of sound may be above the real position. The main disadvantage, however, is the deadness and poor tonal quality of such auditoria, which has forced designers to retreat steadily from the principle of direction from the time of the Salle Pleyel onwards. The most recent concert hall, the de Doelen in Rotterdam, has only rudimentary reflectors, while all recent large music studios, such as those in Hanover (Kuhl and Kath, 1963) and Munich (Struve, 1964), depend entirely on a natural distribution of sound, together with good diffusion, to achieve good musical tone and balance.

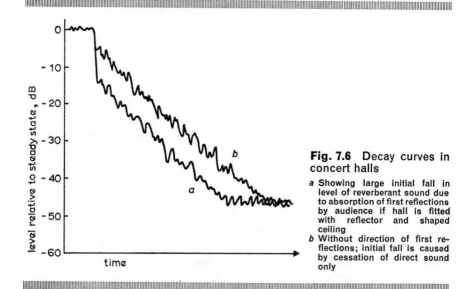

Fig. 7.6 Decay curves in concert halls

a Showing large initial fall in level of reverberant sound due to absorption of first reflections by audience if hall is fitted with reflector and shaped ceiling

b Without direction of first reflections; initial fall is caused by cessation of direct sound only

The methods of geometrical acoustics, nevertheless, permit one to learn a great deal about the distribution of sound in large enclosures, such as large studios and halls. The possibility of the focusing of sound or of echo formation by large surfaces may be studied, as well as the shaping of wall or ceiling areas around stages to assist members of orchestras to hear each other while playing. Use of these methods in designing studios and auditoria has been described in considerable detail by Mankovsky (1971).

7.5 Measurement of reverberation time

7.5.1 General

Most of the commonly used methods of measuring reverberation time depend on the conversion of a signal from a microphone to a rectified voltage proportional to the logarithm of the signal. The direct voltage is then displayed in some manner against a time scale, and the slope of the curve thus produced during the decay of a sound is measured and converted into a reverberation time.

Fig. 7.7 is a block diagram of the equipment. A loudspeaker in the enclosure is fed with a band of random noise (most often one-third of an octave wide) or with a frequency-modulated tone having a depth of modulation of about 5%. The sound from the enclosure is received by a microphone, the output of which is passed, in turn, to bandpass filters for removal of harmonics and noise and then to a logarithmic amplifier and a display device. This is usually a chart recorder capable of recording rapidly varying levels. The lower limit of the reverberation time that can be measured is set by the time constant of the rectifying circuits preceding the display or by the decay time of the bandpass filters.

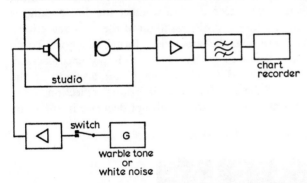

Fig. 7.7 Block diagram of simple equipment for the measurement of reverberation time

The chart recorder is started, and the source of sound is immediately cut off so that the decay of the sound in the enclosure is recorded. The slope of the trace is read by means of a protractor calibrated directly with a reverberation time scale designed to suit the paper speed and the sensitivity of the recorder.

It is often necessary to measure the reverberation time of a studio when it is occupied by, say, an orchestra. For this purpose, the decay of a burst of wideband noise is recorded on a magnetic-tape recorder and replayed through the bandpass filters in sequence into the logarithmic chart recorder.

A sound source often used for this purpose is a blank-cartridge pistol, generally of about 10 mm calibre. This barbarous method has several disadvantages, not the least being the discomfort and shock felt by the more sensitive of the unwilling musicians. The initial sound pressures are high enough to produce nonlinear effects in some sound absorbers, and the results show curious features that often cannot be explained, such as an increase in reverberation time at low frequencies when the orchestra is present. It is better to use any other available method. The orchestra may, for instance, be asked to play a loud chord with component notes spread over a wide frequency range; an organ provides a good wide-range source in the same way. The organ has the advantage, even over the pistol, in the power available at the lowest test frequencies.

The fitting of the best straight line to a decay curve is a matter of human judgment, since the curve may deviate markedly from a straight course. There will be an initial change of slope, such as that shown in Fig. 7.6a, which may take the form of an initial rise or fall of the trace, according to whether the direct sound is in phase or out of phase with the resultant of the early reflections. Deviations may take several characteristic forms, which will be recognised by an experienced observer and their importance assessed. There will be variable fluctuations purely on account of the statistical nature of the summation of individual reflections from the walls. There may be a steady or abrupt decrease in the slope

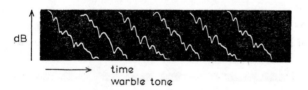

warble tone

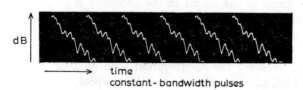

constant-bandwidth pulses

Plate 7.1 Successive decay curves from small studio
a Using short bursts of warble tone
b Using constant-bandwidth pulses

of the decay due to poor diffusion, as will be seen in Chapter 9, or there may be regular beats of large amplitude due to the interference of two closely spaced room modes. When using a test signal of finite bandwidth, derived from random noise or a frequency-modulated tone, each successive decay in a series of replications will be different, so that the accuracy can be improved by averaging the estimated slopes of several decays. Geddes and Gilford (1957) experimented with a finite-bandwidth test pulse, of which the time function remained identical from one pulse to the next; this produced a series of identical decay curves with successive signals, but the variability of measurements from the traces by different observers was greater than that obtained with random sources, because the averaging of several successive but varying traces was no longer possible. Plate 7.1 shows groups of traces obtained at the same frequency with a frequency-modulated tone (*a*) and constant time-function test pulses (*b*).

7.5.2 Direct measurement using cathode-ray oscilloscope

For a large organisation carrying out large numbers of reverberation measurements for research or routine control purposes, it is an advantage to replace the chart recorder by a faster measuring system. Somerville and Gilford (1952) used a logarithmic amplifier coupled to the *Y* plates of a cathode-ray oscilloscope. The measurement chain is shown in Fig. 7.8. The emission of the test signal from the loudspeaker is controlled by a 'tone pulser', the function

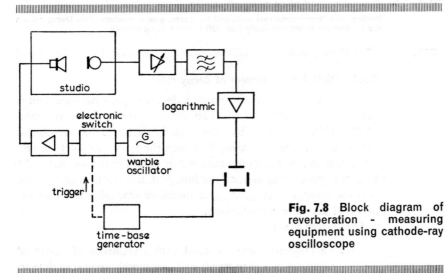

Fig. 7.8 Block diagram of reverberation - measuring equipment using cathode-ray oscilloscope

of which is to interrupt the signal periodically. At the moment of interruption, a triggering signal is sent to the oscilloscope, starting the time base and thus initiating the display of the decay. A suitable logarithmic law is derived from the voltage across a pair of semi-conductor diodes connected back to back and fed from a high-impedance source. The reverberation time is read directly from a graticule arranged to rotate about the axis of the oscilloscope screen. Calibration is effected by displaying a stepped decay and adjusting the time-base speed until the calibration decay lies along a reference line on the graticule.

By driving the system from an oscillator provided with a slow frequency sweep, and photographing the successive traces on a slowly moving film, composite displays, such as that shown in Plate 7.2, could be produced. These, and a variant using coherent detection (Gilford and Greenway, 1956), have proved to be power-ful tools in the investigation of colourations in small studios.

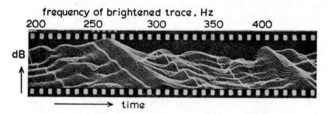

Plate 7.2 Typical pulsed glide of small studio

Vertical axis is pressure-level axis, and horizontal axis represents time. Decay curves are for steadily rising frequency from left to right along display

7.5.3 Digital measurement of decay rates

It has been remarked above that a degree of personal judgment is needed in the fitting of the best straight line to a decay trace. Moffat (1967) of the BBC Research Department has developed methods of digital processing by which the reverberation time and other features of the decay curves can be evaluated mechanically by a data-processing system, the interpretative features being built into the computer program, and therefore amenable to examina-tion and accurate reproduction.

The system consists of

(a) a magnetic tape recorded with a sequence of bursts of noise, ⅓ octave in bandwidth, accurately spaced in time so that precise trigger signals can be derived from the leading edges of the bursts. This tape is reproduced by means of loudspeakers into the studio under test, and the output of a microphone in the studio is recorded on a second tape, which is subsequently processed in the laboratory

(b)   an analogue–digital convertor, with associated circuits, by means of which digitising is started at the beginning of one of the recorded decays, spurious signals from external sources being automatically rejected. The output from the convertor is a punched paper tape

(c)   computer and peripheral equipment that accepts the paper tape and is programmed to print out the reverbera-tion time of the decays from each of 27 noise bands. (In the BBC equipment, the computer is preceded by a transcoder, the purpose of which is to allow an 8-hole tape to be converted to 5-hole tape, in which each sample and a parity check occupy two characters.)

Details of the equipment and the reverberation-time program are given by Moffat (1967) and also by Spring (1971). The computer is programmed, in effect, to

(a)   find the start of the decay

(b)   find the point at which the decay disappears into noise

(c)   compute the mean slope between these two points

(d)   print out the table of reverberation times together with a statement of the number of decibels difference between the start and end of the measured section, and of the standard deviation of the results at each test frequency; any decay that gives a reverberation time far from the mean is printed out as a curve for further inspection.

This method has now been in continuous use for measurements on new studios and rechecks on existing studios for about six years. The prerecorded tapes with the test signals are posted from the BBC Research Department to regional studio centres for the tests and returned for analysis.

The automatic processing of decay curves has also been used for other purposes in studio-acoustics research (Chapter 9).

7.5.4   Schroeder's method of processing decay curves

It was noted in Section 7.5.1 that even decays that were free from the more specific types of deviation from the exponential law were subject to random fluctuations associated with the statistical nature of the summation of the reflections comprising the reverberant sound. Schroeder (1965) has shown that, although these fluctuations will differ from one decay to the next, it is possible to obtain, from an individual decay, a curve that is the ensemble average of all possible decays derived from input pulses with the same spectrum.

He does this by processing the decaying sound of a short tone burst that covers the required spectral band to yield an instantaneous function

$$\int_t^\infty p^2 \, dt$$

where p is the pressure at time t during the decay, and the integration is performed, as indicated, from time t to infinite time.

The derivation of this integral obviously raises problems, because it cannot be carried out in one operation in real time. Instead, the decay may be recorded on tape, and then played backwards into an integrating circuit. Alternatively, as pointed out by Kuttruff and Jusofie (1967–68), we can note that

$$\int_t^\infty p^2 \, dt = \int_0^\infty p^2 \, dt - \int_0^t p^2 \, dt \quad . \quad . \quad . \quad . \quad (7.13)$$

and derive the wanted integral by subtracting a voltage proportional to the second term from one representing the whole integral from zero time to infinity.

The first method is inconvenient, because it involves a series of separate tape manipulations, and the second depends on a very high degree of accuracy and stability in the processing equipment.

Schroeder has applied this method to the analysis of decays from the New York Philharmonic Hall and other auditoria, and demonstrated their freedom from adventitious fluctuations. Apart from this, no systematic use of the method has yet been reported.

7.5.5 Reverberation time from rate of phase change

The methods described above all require the possession of apparatus capable of carrying out measurements on the logarithms of rapidly falling sound pressures. A method using only quasi-steady-state observations has been proposed by Schroeder (1959). If a loudspeaker and a microphone are set up at a known distance d apart from one another, and pure tone of steadily rising frequency is radiated by the loudspeaker, the rate of change of phase angle

δ with frequency f, between the transmitted and received signals, is given by

$$\frac{d\delta}{df} = T/2\cdot2 \quad . \quad . \quad . \quad . \quad . \quad (7.14)$$

All that is necessary, therefore, is to connect the loudspeaker-signal voltage to one pair of plates of an oscilloscope and the microphone signal to the other. Elliptical Lissajous figures are then produced that take all configurations between diagonal lines and circles as the relative phase changes. The frequencies at which the Lissajous figure reaches the same configuration in successive rotations is noted and plotted against successive increments of 2π on a phase-angle scale. This method has not been much exploited, but is worthy of note, since it only requires apparatus that is available in almost every audiofrequency-test laboratory.

The method is not applicable at low frequencies where the modes are separated from each other, since here the phase lag increases by π for each successive normal mode, and is therefore independent of the reverberation time.

7.6   Artificial reverberation

7.6.1   General

An artificial increase in reverberation time is often desirable for adapting a studio for different programmes or to provide varying acoustic perspectives in, say, a drama programme. The reverberation is added either in the studio itself or electrically in the programme output from the studio. These two schemes are represented in Figs. 7.9a and b.

Addition of the reverberation in the studio itself has the advantage that the performers share in the acoustic environment; musicians appreciate this particularly, as it facilitates good ensemble playing and natural unforced production of vocal or instrumental tone. On the other hand, the addition of reverberation in the studio is generally more difficult and expensive than adding it electrically to the output.

7.6.2   Addition of reverberation in studio

The simplest method of increasing the reverberant sound in the studio is to feed the output of a microphone in the studio to a loudspeaker some distance away, so that a delayed feedback is produced. The obvious disadvantage of this is that the loudspeaker

will feed energy back to the microphone, and there may be instability at frequencies for which the fedback sound at the microphone
is in the same phase as the original programme sound. This simple
method is therefore not to be recommended, although the author
has had worthwhile results with it under favourable conditions;
as an example, it was found possible to improve the reverberation
for organ music in a very acoustically dead church by means of
feedback from loudspeakers in positions that were screened from
the microphone.

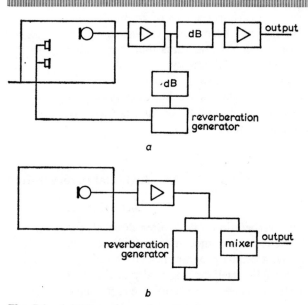

Fig. 7.9   Addition of reverberation to programme
a In studio   *b* In sidechain

Vermeulen (1955) introduced a system of artificial reverberation'
which he originally called 'stereo-reverberation' and which was
later marketed under the name of 'ambiophony'. This system is,
in effect, an extension of the simple feedback just described, in
which chosen delays can be inserted between the microphone and
the loudspeakers by recording the microphone output on a
magnetic-tape loop and replaying it from a series of reproducing
heads spaced out along the length of the loop (Fig. 7.10).
    Just as in the simple feedback system, the principal limitation
lies in the colourations produced by harmonic series of feedback

frequencies, and ultimately in the possibility of actual instability. Some 60 or more loudspeakers are scattered around the walls and ceiling of the studio, and are fed in groups by power amplifiers connected to the several reproducing heads. This provides a very varied set of loop delays, so that no particular path length is associated with more than a small fraction of the total energy.

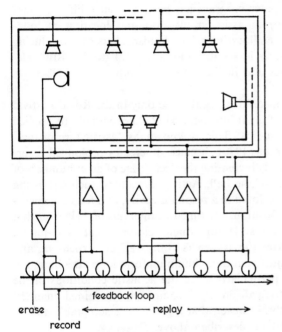

feedback loop

erase

replay

record

Fig. 7.10 Schematic of Ambiophony system

Heads are mounted on circular track, with means for altering individual delays in steps of 15 ms

Nevertheless, the system must be very carefully adjusted if an appreciable increase of reverberation time is to be obtained and if the reverberation is to be free of noticeable colouration or flutters. It is generally necessary to keep the system microphone as near as possible to the performers, so as to reduce the gain required in the reverberation channels, and a directional microphone is an advantage. To obtain a long reverberation time, a long average loop delay must be chosen, and the loop attenuation must be small. If the loop delay is longer than about $\frac{1}{4}$ s, flutters become increasingly noticeable, and the loop attenuation must therefore be low. For example, to obtain a 2s reverberation time, with a loop delay of $\frac{1}{4}$ s, the attenuation round the loop must be $\frac{1}{4} \times 60 \div 2 = 7 \cdot 5 \, \text{dB}$.

It has been shown by Schroeder (1954) that the transmission characteristic of any room between a loudspeaker and a microphone has fluctuations of the order of 10 dB at closely spaced frequencies; therefore, if the system is adjusted to a mean loop attenuation of 7·5 dB, the increase of loop gain required to produce instability at some frequency is only 2·5 dB. The margin for manoeuvre is therefore very small between flutter and instability, and rapidly becomes smaller as the intended reverberation time is increased. A computer program has been devised (Gilford, Jones and Moffat, 1966) to assist in the optimum location of loudspeakers and in the allocation of the loudspeakers to reproducing heads. The system is in regular and successful use for BBC television programmes and in the recording studios of a wellknown recording company.

Another system, so far in regular use only in the Royal Festival Hall, London, is of great interest, and could eventually be applied to broadcasting studios. This is known as 'assisted resonance' (Parkin and Morgan, 1970), because it creates prolonged reverberation by effectively increasing the decay time of great numbers of the natural modes of the hall. Active resonators, each consisting of a loudspeaker enclosed in a resonant cavity, and fed by a microphone in another position, are tuned at intervals of 3 Hz over that part of the musical scale for which reinforcement is required. In its present form, the system is comparatively expensive, and needs periodic adjustment to maintain uniform loop attenuation in all the channels, particularly since the acoustic conditions in the hall are found to drift in an incompletely explained manner. There is little doubt, however, that it gives more acceptable quality than the other systems described above.

7.6.3 Addition of reverberation to studio output

If a pressure pulse of very short duration is radiated into a room, the subsequent time function of the sound pressure will consist of a series of repetitions of that pulse corresponding to reflections of the original pulse from the surfaces of the room. Let us represent this as a function $h(t)$. If we replace the single pulse by a programme sound which we may represent by a time function $g(t)$, the reverberation will modify the sound so that, at any instant, the sound pressure is given by the expression

$$\int_0^\infty g(t)h(t-\tau)\,d\tau \quad . \quad . \quad . \quad . \quad (7.15)$$

which is normally known as the convolution of $h(t)$ with $g(t)$.

The spectrum of this combined function may be shown to be the product of the spectra of the two functions separately, and the characteristic timbre of the function $h(t)$ is heard superimposed on the original programme. The object of almost any artificial reverberation device is to produce a function $h(t)$ which resembles, in all its important characteristics, the impulse response of a real room.

The impulse response of a room is typically as shown in Fig. 7.11. The early reflections of the original sound are widely spaced, and their amplitudes are large. In a large hall, the early spacings will be some tens of milliseconds, but the subsequent spacings rapidly diminish, eventually decreasing inversely as the time after the start. At the same time, the amplitudes diminish at a greater rate, so that the product of the density and individual energies of the reflections eventually vanishes.

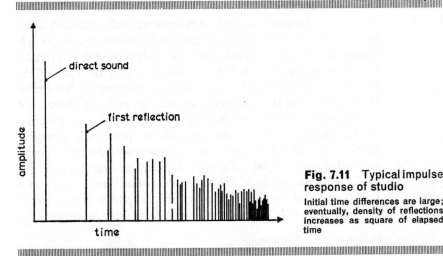

Fig. 7.11 Typical impulse response of studio

Initial time differences are large; eventually, density of reflections increases as square of elapsed time

The simplest reverberation device consists of a delay line with feedback to the imput, which has adjustable attenuation. This is analogous to the single-loudspeaker system described in Section 7.6.2. Examples of these delay-line systems are single-coil springs, simple magnetic-tape loops with spaced recording and reproducing heads and airpath delays. The impulse response of such a system is given by

$$h(t) = \delta t + g\delta(t - \tau) + g^2\delta(t - 2\tau) + \dots + g^n\delta(t - n\tau) + \dots \quad (7.16)$$

where $\delta(t)$ is a unit impulse occurring at time t, τ is the delay time and g is the attenuation round the delay path.

The spectrum of this function is

$$e^{-\omega\tau}+ge^{-2\omega\tau}+g^2e^{-2\omega\tau}+... \quad . \quad . \quad . \quad (7.17)$$

This clearly consists of a set of line frequencies, multiples of $1/2\pi\tau$. A device such as this imposes on the programme a characteristic 'comb-filter' colouration, which sounds like the sound of a hand-clap between two parallel walls in the open air. The programme can, of course, be passed through several such systems with different delay times to avoid having a single strong colouration of this sort, but this proves to be useless, because the several characteristic colourations can be heard independently, and the final result is no better. Schroeder and Logan (1961) have shown, however, that the addition of feedback in specific relation to the gain of the main chain (Fig. 7.12) converts the comb filter with variable amplitude into a spectrum of constant amplitude but periodically varying phase. For a single system, the colouration is still clearly audible. If, however, several such systems with unrelated delay times are connected in cascade, virtually colourless reverberation is obtained. The use of this principle is severely restricted, in practice, for two reasons. First, it is not possible to alter the ratio of the reverberant to the direct sound without upsetting the fixed relationship between the amplitudes in the main and side chains, and consequently introducing amplitude variations in the spectrum of the output. Secondly, the necessity to pass the programme through several delay circuits in cascade makes the achievement of a satisfactory signal/noise ratio difficult, if not impossible.

The most successful of the simple delay-line types consists of a helical spring along which the programme is passed in the form of torsional waves in the wire of which the spring is wound. An

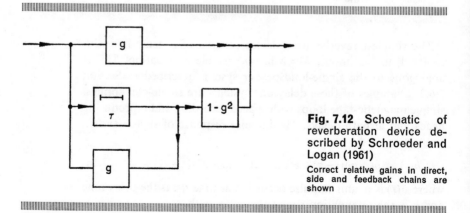

Fig. 7.12 Schematic of reverberation device described by Schroeder and Logan (1961)

Correct relative gains in direct, side and feedback chains are shown analysis of the behaviour of such springs was given by Meinema, Johnson and Laube (1961). Generally, in the best equipments, two springs with slightly different delay times are used in parallel in the same electromechanical system, $1/\tau$ being of the order of 12–15 Hz. These values are too low in pitch to be heard as fundamentals of a comb-filter tone, and, for most purposes, a device of this type gives very acceptable reverberation. It is marred by the occurrence of some apparently unavoidable axial modes of vibration, which are heard as colourations at several frequencies. They will be seen as gently sloping traces in the pulsed-glide display in Plate 7.3. These colourations render the equipment unsuitable for use with such programmes as symphonic orchestral concerts, which are subject to sensitive aesthetic criticism.

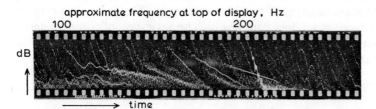

Plate 7.3 Pulsed glide of helical-spring reverberation device, showing long decays between 100 and 200 Hz caused by longitudinal transmission

Another method that has been used in as extension of the simple magnetic-tape system described above, in which several reproducing heads are used and fed back to two or more recording heads. This type has not been outstandingly successful in spite of development to a high pitch of sophistication (Axon, Gilford and Shorter, 1955).

The most successful methods, in terms of suitability for programmes of high artistic quality, are those in which the colourations associated with combinations of a few simple delays with feedback have been avoided by the use of 2- or 3-dimensional delay systems which introduce irrationally related series of delays.

The oldest method, and still in some respects the best, is to radiate the programme from a loudspeaker into a room that has the desired reverberation time. The output of a microphone in the room then consists of the original programme convoluted with the impulse response of the room. With some disadvantages, which will be referred to below, such 'echo rooms' can produce acceptable results for a wide variety of programmes.

Echo rooms are now being largely supplanted by the reverberation plate devised by Kuhl (1958). This consists of a rectangular sheet of steel, 0·5mm thick, measuring 1m × 2m in area, and suspended from spring mountings at its four corners. The sheet is carefully selected at the rolling mill to have no measurable double curvatures, since these would greatly increase its bending stiffness. The coil of a moving-coil loudspeaker unit is attached to the plate, and a pot magnet with a 3mm gap is fixed to the frame to register with the coil. Programme-signal currents, passed through the coil, set the plate into bending vibration, and the vibrations are picked up at another point on the plate by a piezoelectric accelerometer attached to it. The steel sheet acts as a 2-dimensional delay system. No feedback is necessary, because the sheet itself has a free decay time up to about 10s at low frequencies. The decay time may be reduced at will by a control which brings a sheet of porous material into close proximity with the steel sheet, thereby providing resistive damping to the latter.

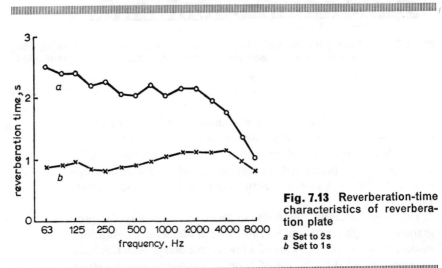

Fig. 7.13 Reverberation-time characteristics of reverberation plate
a Set to 2s
b Set to 1s

Fig. 7.13 shows the reverberation characteristics obtainable with different settings of the porous sheet.

One of the defects of a small echo room is that the lower room modes are widely spaced in frequency, and therefore the bass end of the frequency scale tends to be coloured. The reverberation plate, on the other hand, vibrates in bending, and the velocity of the bending waves is proportional to the square root of the frequency. At low frequencies, therefore, the velocity is low, and the modal frequencies are low and closely spaced. The low-frequency reverberation is therefore clean and free of colourations. At higher frequencies, colourations can be heard, causing the reverberant sound to be slightly metallic in quality.

Table 7.2 summarises the good and bad features of common artificial-reverberation systems. There is no perfect system, if this means one that is indistinguishable from the genuine reverberation of a good studio; the use of a large studio as an auxiliary reverberation room is artistically the most satisfactory method, but it is usually impracticable.

Table 7.2 Advantages and disadvantages of artificial reverberation systems

System	Advantages	Disadvantages
Ambiophony	Reverberation present in studio	Colourations and flutters limit performance Comparative size and expense
Assisted resonance	Absence of colourations	Expense and complication Continuous supervision
Magnetic tape or drum	Simplicity, compactness	Colourations and flutters
Schroeder–Logan	Uncoloured if several delays in cascade	Indifferent signal/noise ratio Not easily amenable to direct/reverberant-ratio adjustment
Reverberation room	Good quality in middle and high frequencies Simplicity Easy adjustment of reverberation	Small-room sound due to widely spaced modes Low-frequency colouration Large volume
Kuhl plate	Clean bass Smaller and cheaper than reverberation room Instant adjustment of reverberation time	Metallic colourations around 1kHz
Helical springs	Cheapness Portability, small size	Colourations

Chapter 8

Sound absorbers

8.1 General

The sound-absorption coefficient of a material has been defined earlier, and it has been shown that the accurate knowledge of this property for all materials used in the construction of a studio is necessary for the correct prediction of its acoustic behaviour. Methods of measuring the absorption coefficient are now described, followed by a description of sound absorbers of various types and their functions in studio treatment.

8.2 Measurement of absorption coefficients

There are two methods in common use for measuring absorption coefficients. They are known as the standing-wave-tube method and the reverberation method.

8.2.1 Standing-wave-tube method

This method is best applied to the measurement of the absorption coefficients of materials at normal incidence, using small samples with linear dimensions of a few centimetres. The apparatus (Fig. 8.1) consists of a tube, typically about 1m long, one end of which is closed by a removable rigid cap. The other is connected to a transducer, normally a conventional loudspeaker, by which tone from an oscillator can be injected into the tube, forming a system of standing waves by interference of the injected sound with sound reflected from the termination of the tube. A sample of the same diameter as the interior of the tube is cut from the test material and inserted in the end of the tube in front of the end cap. The standing-wave ratio of the sound field is then measured by means of a probe microphone, the orifice of which can be moved along the axis of the tube. If n is the ratio of the maximum pressure to the minimum pressure along the tube, the normal-incidence absorption coefficient can be shown to be

$$N = 4/(n+1/n+2)$$

Fig. 8.2 shows a graph of the absorption coefficient against n.

A full account of the method, with details of the operations and corrections required, have been given by Beranek (1950).

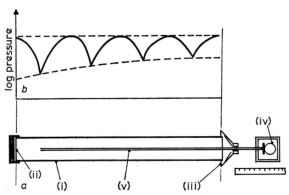

Fig. 8.1 Standing-wave tube for measurement of absorption coefficients
    a Section
       (i) tube
      (ii) sample
     (iii) loudspeaker
     (iv) microphone
      (v) probe tube
    b Variation of pressure along tube

If the exact distance of the first standing-wave minimum from the face of the sample is also measured, the real and imaginary parts of the acoustic impedance at the front of the sample may be calculated, and from these it is possible, in many cases, to determine the random-incidence absorption coefficient which corresponds to the conditions obtaining in the normal use of the material. Methods for the calculation of the impedance are given by Atal (1959) and Dubout and Davern (1959).

The tube method, since it requires only very small samples of the material, is a very useful technique for the study of absorbing materials and constructions during the process of development. However, the calculation of the random-incidence absorption

F

146 *Sound absorbers*

depends on the assumption that the impedance remains constant at all angles of incidence. Although this is approximately true for a very large proportion of absorbers, there is an element of doubt that makes it essential to carry out direct determinations of random-incidence coefficients with large samples on every material that is to be used for studio treatment. This may be done by the reverberation method.

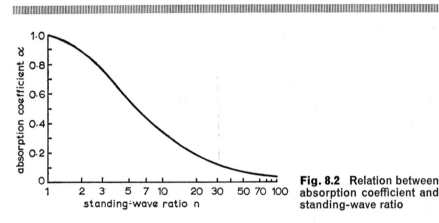

Fig. 8.2 Relation between absorption coefficient and standing-wave ratio

8.2.2 Reverberation method

The principle of the reverberation method is to measure the reverberation time of a room by the methods described in Chapter 7 and then to repeat the measurement after installing a known area of the experimental material on a wall or other surface of the room. The total absorption in the room is calculated from each measurement by Eyring's formula, and the contribution of the sample is found by simple subtraction. BS3638:1963 specifies the properties of the 'reverberation room' that should be used for these measurements. The volume of the room should be 180–$250\,\text{m}^3$; in rooms smaller than the lower limit, the modal frequencies are too widely spaced to give reliable results below $125\,\text{Hz}$. All surfaces of the room should be faced with hard sound-reflecting finishes, and the reverberation times should be not less than those shown in Table 8.1.

The source of sound is a loudspeaker placed in a position where it will excite a large number of room modes; a suitable position is at one-third of the distance along a diagonal of the room between two opposite corners. It is an advantage to have two loudspeakers,

Table 8.1   Minimum reverberation time of reverberation room (from BS3638:1963)

Frequency	Reverberation time
Hz	s
63–500	5
1000	4
2000	3
4000	2
8000	1

which are used alternately, and the reverberation time should be measured in at least five different microphone positions for each loudspeaker position. The most rapid results are obtained if the ten combinations of the two loudspeakers and five microphones can be selected sequentially by a rotary switch and the readings of reverberation time and added averaged on a desk calculator as they are obtained.

The size and disposition of the sample is of great importance. BS3638 requires a single sample, $10m^2$ in area, placed off centre on one surface of the room. However, the effect of a single sample will be to attenuate rapidly all room modes having components of particle velocity perpendicular to the treated surface, and leaving the other modes comparatively unaffected. The measured reverberation time, being largely influenced by the decay of the long-lasting unattenuated modes, will yield low values of the absorption coefficient. BS3638 therefore further requires the provision of scattering surfaces, such as hanging sheets of hardboard, to deflect the wave fronts, so that sound initially travelling parallel to the absorbing surfaces is continually changed in direction.

The same result as shown in Chapter 9 can be obtained by subdividing the sample into three or four smaller areas on different room surfaces. However, this also causes a marked increase in the apparent coefficient of the sample at middle frequencies. An absorbing area in a reflecting wall causes diffraction of the sound waves, because the lines of flow of the particles of air converge towards the absorber. Additional sound energy that would otherwise be reflected from the surrounding walls therefore falls on the absorber and is dissipated. If the sample is subdivided, this anomalous absorption is increased, and thus a subdivided sample gives higher measured coefficients than a single sample of the same total area. Fig. 8.3 shows the effects of the scattering of the sound field and the subdivision of the sample. Curve *a* is the absorption coefficient of a rockwool material, calculated from impedance measurements in a standing-wave tube, and therefore representing approximately the random-incidence coefficient of an infinite-area sample of the material. Curve *b* was measured in a reverberation room using a single 10 m² sample of the same material. Curve *c* shows the result of subdividing the sample into four pieces, each on a different surface, and hanging a number of scattering reflectors in the room.

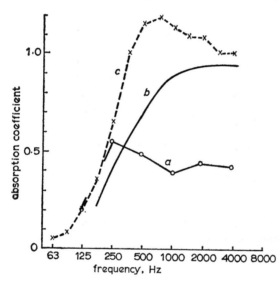

Fig. 8.3 Dependence of measured absorption coefficient of rockwool on distribution and diffusion

a Calculated from impedance-tube measurements
b Single sample; no diffusers
c Divided sample with diffusers

The method adopted for BS3638 was found by Kosten (1959) to give the lowest variation between the results from the same sample measured in different laboratories throughout the world, and probably gives the best representation of the performance of a material when used in large areas for noise reduction in working spaces. However, the divided sample gives a more accurate prediction of the performance of an absorber when used as part of a studio treatment, because the requirements of diffusion and accuracy of design dictate a similar distribution over the walls and ceiling. For this reason, it is the practice in the BBC to use the divided-sample method for all measurements on materials and absorbing constructions to be used in studios. A correction must be made if the absorber is to be used in large continuous areas as in television studios.

8.3 Main types of sound absorber

8.3.1 General

Sound is absorbed by the dissipation of energy in the vibration of the building structures and by viscous loss in the penetration of sound into porous surfaces. Even a fullbrick wall (220 mm thick) has an effective absorption coefficient of about 0·05 at 100 Hz, and that of plasterboard on wood framing or of plastered clinker blocks may be as high as 0·4 at this frequency. Wood panelling, doors, windows and wooden floors behave similarly. The surfaces of most common building materials also provide an appreciable amount of porous absorption, and essential furniture and floor coverings must also be taken into account.

The third mechanism of absorption is that encountered by sound in its passage through the air. This is unimportant below 1 kHz, but increases steeply with frequency. In large concert halls and studios, where the mean free path between reflections is long, air absorption becomes relatively important, and may account for more than half the total energy loss at 8 kHz.

The remainder of this chapter will deal with absorbers classified according to the frequency range over which they operate most usefully. Porous absorbers are mainly used for middle- and upper-frequency absorption, and, as will be seen in more detail below, a porous layer absorbs most effectively when its thickness is comparable with a wavelength. For low-frequency sounds, the thickness necessary to satisfy this condition is prohibitive, and more satisfactory results can be obtained by the use of systems which are excited into resonance by the sound and dissipate its energy by internal losses.

8.3.2 Low-frequency sound absorbers: Helmholtz resonators

Two types of resonant system are in common use as sound absorbers: membranes or panels and Helmholtz resonators.

The Helmholtz resonator is the simplest type. It consists of a volume of air enclosed by a rigid container with a hole or neck by which the contained air communicates with the exterior. The mass of the plug of air in the neck and the compliance of the contained air volume together constitute a resonant system of which the resonance frequency is given by Rayleigh (1896) for the case of a cylindrical neck.

$$f = \frac{c}{2\pi} \sqrt{\frac{A}{l'V}} \qquad \cdots \cdots \quad (8.1)$$

where

c = velocity of sound in air
A = cross-sectional area of neck
V = volume of contained air
l' = effective length of neck $= l + \pi r/2$
l = actual length of neck
r = radius of neck

Another common form of resonator consists of a cavity of uniform cross-section with a long slit running along its length. In this case, the frequency is given by an implicit equation (Pederson, 1940).

$$f = \frac{c}{2\pi} \sqrt{\left[\frac{b}{\{l + 0.5b + (2/\pi) \log_e(\lambda/\pi b)\}A'} \right]}$$

where

A' = cross-sectional area of the cavity
b = width of the slit
l = depth of the slit
λ = wavelength of sound at the resonance frequency

In the neighbourhood of the resonance frequency, the velocity of the air in the neck or slit is high, and dissipation of sound energy takes place as a result of viscous friction. The resistance of the neck may be raised to a suitable value by covering it with gauze or cloth or by plugging it with a fibrous material.

A careful examination of the absorbing action of the Helmholtz resonator is justified, even in the space of this short monograph, because it provides a clear understanding of the roles of internal and radiation resistance, 'pressure doubling', resonance and re-radiation without which the bandwidth, peak absorption, scattering and energy storage cannot be explained. With this understanding, it is possible to predict, at any rate qualitatively, the properties of all other forms of absorber.

Fig. 8.4 represents a resonator of volume V_0 that lies behind a hard wall having a cylindrical opening of frontal area A. Let the sound pressure at the opening of the neck be $p(t)$, where t is the time variable. The forces on the air in the neck in an upward direction are

(a)   the force $-Ap(t)$ due to the external sound field

(b)   the frictional force $-Bz(t)$ due to neck resistance, where $z(t)$ is the upward displacement of the air in the neck from its equilibrium position

(c)   the elastic force $-Yz(t)$ due to the contained air. B and Y are constants related to the frictional and elastic properties of the air, respectively.

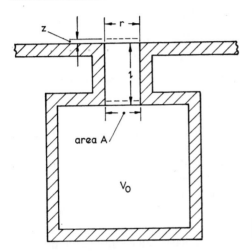

Fig. 8.4 Section through Helmholtz absorber

The equation of motion of the plug of air in the neck is therefore

$$Al\rho z + Bz + Yz + Ap = 0 \quad \cdot \ \cdot \ \cdot \quad (8.3)$$

where l is the length of the neck and ρ is the density of the air and writing p for $p(t)$ and z for $z(t)$.

Assuming adiabatic compression and expansion of the air, we have, from the well-known gas laws,

$$PV^\gamma = \text{constant}$$

where γ is the ratio of the specific heats of air at constant pressure and V is the volume of a given mass of gas at that same pressure. Differentiating,

$$dP/P + \gamma dV/V = 0$$

Now Y is the constant of proportionality between the force on the plug of air in the neck and the deflection z.

The force on the neck, however, is AdP, the change in volume dV is Az, and therefore

$$Y = \frac{A^2 dP}{dV} = \frac{A^2 \gamma P}{V}$$

But

$$\gamma P = c^2 \rho$$

(Chapter 1). Hence

$$Y = \frac{A^2 \rho c^2}{V} \quad . \quad . \quad . \quad . \quad . \quad (8.4)$$

The equation of motion of the air in the neck therefore becomes

$$\frac{Al\rho \ddot{z} + B\dot{z} + c^2 \rho z A^2 + Ap = 0}{V_0} \quad . \quad . \quad . \quad (8.5)$$

Now p consists of two components, and may be written

$$p = p_i + p_r$$

where p_i is the pressure of the incident sound and p_r is the reaction of the external air to the movement of the air in the neck. The surface of the air in the neck, as it oscillates, radiates sound into the surrounding air, and the pressure set up on the radiating surface for unit volume flow is known as the radiation impedance, which is a function of the geometry of the surface, the frequency of vibration and the nature of the medium into which it radiates.

The radiation impedance is, in general, a complex quantity with real and imaginary parts, which we will call R_r and X_r, respectively, representing the inphase and quadrature components of the pressure, respectively, with respect to the velocity of the radiating surface.

The equation of motion (eqn. 8.5) will be seen to be exactly the same as that for a series resonant circuit if we make z represent current, p voltage and the constant terms the several impedances. The equivalent circuit is then as shown in Fig. 8.5. The driving

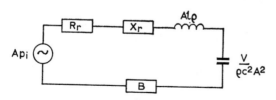

Fig. 8.5 Equivalent circuit of Helmholtz absorber

voltage is Ap_i, R_r is the radiation resistance and X_i is the radiation reactance, the other lumped components being the coefficients in the equation of motion.

For small resistances, the resonance frequency is seen to be

$$(1/2\pi)\sqrt{\{(V_0/\rho c^2 A^2)(Al\rho + Y_R)\}} \quad . \quad . \quad . \quad (8.6)$$

the radiation reactance being, here, a small inertance, which is usually regarded as an end correction, slightly increasing the length of the neck to the effective value l' (eqn. 8.1).

Hence we may write

$$f = (c/2\pi)\sqrt{(A/l'V)}$$

which is the result quoted above.

Let us now consider the absorption of sound by the resonator at its resonance frequency. The reactive components will have cancelled each other out, leaving only resistances in the circuit.

The volume flow in the resonator is then given by

$$\frac{Ap_i}{R_r + B}$$

and hence the rate of dissipation of energy in the neck of the resonator is given by

$$\frac{(Ap_i)^2 B}{(R_r + B)^2}$$

and, if we put $B = \mu R_r$, this becomes

$$\frac{(Ap_i)^2 \mu}{R_r(1+\mu)^2} \quad . \quad . \quad . \quad . \quad . \quad (8.7)$$

which reaches a maximum value of $Ap_i^2/(4R_r)$ when $\mu = 1$. This is clearly analogous to the matching of an electrical generator to a load, the load taking the greatest power when its resistance is equal to that of the generator driving it.

Now the pressure p_i, shown as the 'generator voltage' in this circuit, is, according to Thévenin's theorem, the open-circuit pressure, i.e. that which would exist if the resonator were replaced by an infinite impedance. It is therefore twice the pressure in the incident wave, because the pressure would be doubled if the resonator were replaced by a hard surface, which would be the analogue of Thévenin's infinite impedance. We may now calculate the total-energy-absorption rate from a sound field of known intensity, and

F*

from this we may derive the effective absorbing cross-section of the resonator. This is defined as the area of plane wave, falling normally on the surface containing the resonator, that carries an energy flow equal to that absorbed by the resonator.

It may be shown (Crandall, 1926) that the radiation resistance of a piston moving in a small hole in an infinite plane is given by $2\pi\rho c/\lambda^2$, where λ is the wavelength of the sound.

Hence, substituting in expr. 8.7, the rate of dissipation of energy in the neck of the resonator is

$$(Ap_i)^2\lambda^2\mu/\{2\pi c(1+\mu)^2\} \quad . \quad . \quad . \quad . \quad (8.8)$$

A plane wave of area G and pressure $p_i/2$ would be conveying energy at the rate of $Gp_i^2/4\rho c$, and therefore, if we equate this to the dissipation in expr. 8.8, we find that

$$G = 2A^2\lambda^2\mu/\{\pi(1+\mu)^2\} \quad . \quad . \quad . \quad . \quad (8.9)$$

which has a maximum value of $A^2\lambda^2/2\pi$ for correct matching when $B = R_r$.

It might appear, at first sight, that there is a dissipation of an equal amount in the 'radiation resistance', and this has been incorrectly described elsewhere as being a result of scattered radiation in the form of spherical waves. But this is clearly not necessarily so, because the same argument, using an equivalent circuit, could be applied equally to the absorption at an arbitrary plane drawn across a uniform duct down which sound was passing as plane waves. Here the radiation resistance is the characteristic resistance of plane waves in air, namely ρc, and the resistance looking forward could be made to assume the same value by means of a suitable termination. The radiation resistance is analogous to the iterative resistance of a cable or waveguide; and, if we go back along the tube in the direction from which the sound is coming, there will be no dissipation of energy until the source of sound is reached.

On the other hand, if the absorption of sound at the resonator is incomplete, the unabsorbed sound will be scattered. The energy scattered in unit time can easily be shown to be

$$\text{scattered energy} = \frac{Ap_i^2}{R_r}\frac{(1-\mu)^2}{(1+\mu)} \quad . \quad . \quad . \quad (8.10)$$

The fact that, for $\mu \neq 1$, the resonator scatters energy in the form of spherical radiation is of considerable importance (Chapter 9).

A related effect is observed in resonators in which the internal resistance is low, giving a high value of the Q factor. Rschevkin (1938) has shown that, if a Helmholtz resonator of volume V and absorbing cross-section A is placed in a room with volume V' and total absorption A', the reverberation time of the room is increased by a factor

$$\frac{(V' + Q^2 V)A'}{(A' + A)V'} \quad \cdots \quad \cdots \quad (8.11)$$

owing to energy storage in the resonator.

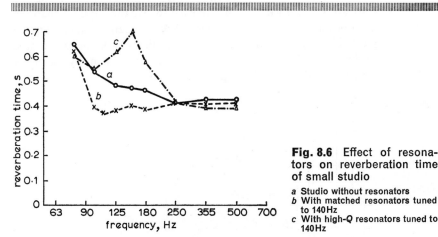

Fig. 8.6 Effect of resonators on reverberation time of small studio

a Studio without resonators
b With matched resonators tuned to 140 Hz
c With high-Q resonators tuned to 140 Hz

For high Q factors, the increase in reverberation time at the resonance frequency may be very appreciable, as shown by Fig. 8.6, which illustrates the effect on the reverberation time of a room without resonators (curve a), with matched resonators (curve b), and resonators with a high Q factor (curve c).

The singular property of Helmholtz resonators, compared with other low-frequency absorbers, is their flexibility. It will be appreciated, from eqn. 8.1, that the frequency of resonance may be varied over a wide range by suitable choice of the volume and of the length and cross-section of the neck. The Q factor, and thus the bandwidth, can be fixed by the choice of the relative values of the mass of the air in the neck and the total resistance.

Now we have

$$\omega = c\sqrt{(A/l'V)}$$

and

$$Q = \frac{\omega A l' \rho}{B} \quad \cdots \quad \cdots \quad (8.12)$$

Hence the half-power bandwidth is

$$\frac{\omega}{Q} = \frac{B}{A l' \rho} \quad \cdots \quad \cdots \quad (8.13)$$

For the maximum coefficient,

$$B = 2\pi\rho c/\lambda^2 \quad \cdots \quad \cdots \quad (8.14)$$

These three equations determine the constants A, V, l' of the resonator, and B will then be adjusted to give the required shape of absorption characteristic on test. The resonance curve may be plotted by inserting the end of a probe microphone through a hole in the back of the resonator and exciting the system by means of tone radiated from a loudspeaker.

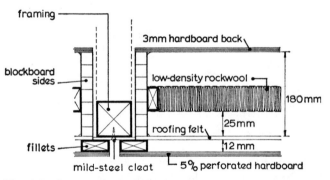

Fig. 8.7 Section through roofing-felt membrane absorber

a Membrane   d Wood framing
b Perforated hardboard cover e Light-density rockwool
c Cleat   f Hardboard or wood back

Resonators may be designed either to damp out general reverberation over a wide frequency band or to suppress isolated modes in small rooms by designing them to have high Q factors and turning them precisely. The theory of their use for the latter purpose has been given by van Leeuwen (1960) of the Nederlands Radio Unie.

Plate 8.1 is a photograph of a studio in which resonators have been applied for wideband absorption; they consist of hollow plaster castings, each divided into several separate compartments

along the length. Each separate compartment communicates with
the air outside through a separate hole. Line arrays such as these
have higher radiation resistance per hole than isolated resonators;
Ward (1952) showed the value to be

$$\sqrt{(2\pi)}\rho c/\lambda d \quad . \quad . \quad . \quad . \quad . \quad (8.15)$$

(where d is distance between successive necks) and gave methods of
predicting the total absorption of such an array. 2-dimensional
square arrays give higher radiation resistance still; this case has
been treated by the author (Gilford, 1952) and others.

Plate 8.1   Broadcasting studio with Helmholtz absorbers
Swansea studio 1

8.3.3 Membrane and panel absorbers

The use of wood panelling for the absorption of low-frequency
sound preceded the scientific study of acoustics, since the control
of bass reverberation in the music rooms of stately homes and the
19th-century concert halls was ensured by the liberal use of panelling. It was also the main source of low-frequency absorption in

early recording and broadcasting studios, and remained so until 1947, when C. W. Goyder used sheets of linoleum stretched over shallow airspaces as low-frequency absorbers in the studios of All-India Radio in Delhi. These absorbed strongly over a narrow band surrounding the resonance frequency determined by the mass of the linoleum and the stiffness of the enclosed air.

Bituminous roofing felt has more suitable rheological properties than linoleum (Gilford, 1952–53). This form of absorber was developed to become the sole low-frequency absorber in the BBC radio and television studios. Fig. 8.7 shows the construction of a typical unit. The resonance frequency and absorption characteristics can be calculated most easily by means of energy equations. Unlike the Helmholtz resonator, the membrane absorber cannot be treated as having a piston-like action, and the displacement z of eqn. 8.5 is a function of the position on the membrane. The resonance frequency of a limp rectangular membrane of sides l_1 and l_2, vibrating in a mode that subdivides the membrane into rectangles with sides l_1/n and l_2/m, can be written as

$$f_{m1}n = (4/\pi^3 mn)\sqrt{(\gamma P l_1 l_2/\sigma V)} \quad . \quad . \quad . \quad (8.16)$$

In practical membranes, the inherent stiffness of the membrane shifts the resonance frequency upwards, the shift being greatest in deep units in which the air-cushion compliance is high.

A sinusoidal distribution of amplitude over the surface of the absorber was assumed in both principal directions, with the possibility that the panel could vibrate in any of several modes thus formed, the lowest being that in which all parts of the membrane move in phase ($m = n = 1$). It was shown that the panel would be expected to absorb strongly in only its lowest modes, others being only weakly excited. However, it appears from experiment that other modes may be excited in certain modular shapes, and therefore units of different shapes and sizes may differ in their effectiveness. This has, on the whole, not presented a serious problem, but nearly square shapes are to be preferred.

A development of the roofing-felt absorber was the 'bonded absorber' patented by Burd and Gilford (1958). In this, the roofing-felt membrane is replaced by a sheet of hardboard, to the back or rough side of which is cemented a sheet of bituminous roofing felt, using a nonhardening adhesive such as those consisting of polyvinyl-acetate emulsions. The composite sheet is fixed over a closed airspace between 50 mm and 300 mm in depth; it behaves generally in the same manner as the plain roofing-felt membrane, but has the advantages of presenting a hard surface to the front which can be painted, and of being rather cheaper to construct.

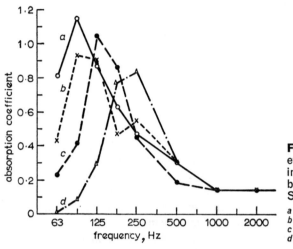

Fig. 8.8 Absorption co-efficient of single-ply roof-ing-felt membrane absor-bers (Burd, Gilford and Spring, 1966)

a Depth = 300 mm
b Depth = 150 mm
c Depth = 75 mm
d Depth = 25 mm

In both of these types of absorber, the resonance frequency can be varied by the choice of the mass of the roofing-felt and the depth of airspace. The roofing-felt absorber can be varied in this manner from about 60 Hz up to 200 Hz peak absorption frequency, but the

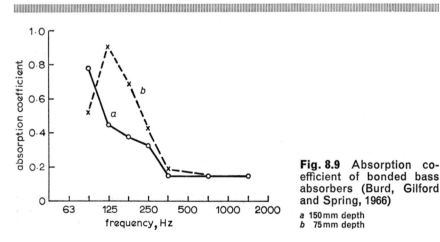

Fig. 8.9 Absorption co-efficient of bonded bass absorbers (Burd, Gilford and Spring, 1966)

a 150 mm depth
b 75 mm depth

bonded absorber, by reason of the stiffness of the hardboard, cannot easily be tuned below 70 Hz. Figs. 8.8. and 8.9 show families of absorption curves for the two types of absorber constructed in different depths, measured as distributed samples in a reverberation room. It will be noted that the absorption coefficient, as based on the frontal area of the sample, can exceed unity. This is because the radiation resistance of the sample, which is considerably smaller than a wavelength, is lower than that for an equal area of an infinitely large sample. [This is fully dealt with in the paper by Burd and Gilford (1958).] As the resonance frequency of the absorber rises with a decrease of depth or membrane mass, the radiation resistance of a sample of fixed area increases, and, to obtain good matching at the higher frequencies, it is necessary to introduce, behind the membrane or the bonded panel, a layer of rockwool or other resistive material. In practice, it is best to do this in any case, as the consequent overmatching in low-frequency absorbers broadens the bandwidth without very much affecting the peak absorption, and thus gives a better product of absorption and bandwidth than does a plain membrane.

Absorbers such as those described above are highly suitable for broadcasting studios, especially those requiring a low-frequency reverberation time that is low for the volume of the studio. In places requiring less efficient bass absorbers, wood panelling is still an effective and economical material. In the large music studios in Hanover (Kuhl and Kath, 1963) and Munich (Struve, 1964), wood panelling has been used, backed by airspaces of three different depths. The spaces are filled with loosely laid rockwool or other similar porous material, not to provide extra damping, but to reduce the natural resonance frequency. As Kuhl points out,

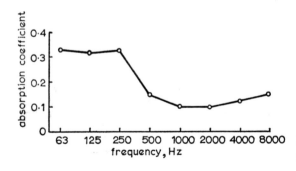

Fig. 8.10 Absorption coefficient of 13 mm wood panelling on 25 mm battens

the stiffness of the airspace is greater if the compressions and rarefactions of the air in the boxes are adiabatic than if they are isothermal, by a factor γ; the function of the rockwool filling is therefore to act as a thermal capacity to reduce the amplitude of the temperature fluctuations, and thus to reduce the stiffness of the air and lower the resonance frequency.

Fig. 8.10 shows the absorption coefficient of typical wood panelling used in this manner.

8.4 Porous absorbers

Sound absorbers consisting of sheets or titles of porous material are familiar in noise-reduction applications, and are indeed the only absorbers widely manufactured for sale. They vary in thickness from about 19 mm to 38 mm, and are designed to be mounted by direct adhesion to the ceilings of rooms or workshops or on frameworks or battens by which they are held at a distance from the hard surface of a wall or ceiling. The front surface of the material may be pierced with holes penetrating nearly to the rear or may be covered with a sheet of perforated metal or hardboard, which can be painted without interfering with the access of sound to the interior of the material. In broadcasting studios, porous materials such as fabrics, glasswool, rockwool fabrics or synthetic wadding are similarly used for the absorption of sound at middle or high frequencies, although they are usually assembled on site with framings and covers instead of being supplied as prefabricated units. The onsite assembly makes provision for subsequent adjustment easier to arrange, should alteration or adjustment of the reverberation characteristic of the room be desirable.

Consider the case of a layer of absorbing material in contact with a hard wall. An incident sound wave travels through the surface boundary with some attenuation, and, after traversing the thickness of the material, is reflected from the hard surface behind. The reflected wave, when it reaches the surface, has a lower amplitude than the incident wave, and therefore a standing-wave system is formed near the wall with a ratio of $(r+1)/(r-1)$, where r is the reflection coefficient.

At the wall itself, the particle velocity is zero, but it increases to a first maximum at a distance from the wall equal to one-quarter of a wavelength.

To obtain the greatest dissipation of energy therefore, the

material should be at least as thick as one-quarter of a wavelength of the sound in the material. For normal-incidence sound, for example, a thin fabric will absorb most energy when it is held at exactly one-quarter of a wavelength in front of the wall, or, to put it in another way, if the fabric is held at a fixed distance, there will be a peak of absorption at every frequency for which the fixed distance is an odd number of quarter wavelengths. The curve of absorption against frequency is therefore as shown in Fig. 8.11, curve *a*.

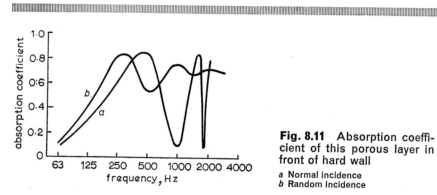

Fig. 8.11 Absorption coefficient of this porous layer in front of hard wall
a Normal incidence
b Random incidence

For random-incidence sound, however, the first maximum is at a frequency for which the distance is one-eighth of a wavelength (Koyasu, 1958), because, at this distance, the mean velocity of sound from all angles from the normal is greatest. Furthermore, the first peak will be broader than for the normal-incidence case, and the subsequent fluctuations will diminish in amplitude, giving an absorption characteristic as shown in Fig. 8.11, curve *b*.

The calculation of the absorption of sound by a layer of porous material of finite thickness is more complicated, and is best carried out by the use of the concept of impedance, as described above in connection with low-frequency absorbers. The reader is referred to more specialised works for details of these calculations (Zwikker and Kosten, 1949), since the existing theory is of little practical utility, giving only approximate predictions.

The most effective method of absorbing sound over a wide range of frequency is by the use of graduated absorbers, which present a continuously increasing resistance to flow as the sound wave progresses through the depth of the absorber. It may take the form of layers of fabric of increasing density and closer packing,

Fig. 8.12 Section through porous wedge absorber
(Dimensions are in millimetres)

or it may consist of cones, pyramids or wedges of porous material with their points or edges facing the incident sound. The theory of graduated absorbers has been given by Head (1965). Fig. 8.12 shows the construction of polyurethane wedges used to cover the walls, floor and ceiling of a free-field sound-measurement room recently constructed by the BBC, and Plate 8.2 is a photograph taken inside the room.

Plate 8.2 Wedge absorbers of polyurethane installed in free-field room

8.5   Practical forms of absorber for studio use

A comparatively small number of types of absorber is sufficient for all studio treatment and construction. The absorbing units in normal use by the BBC will now be described.

The absorption coefficients of all these absorbers are listed in detail by Burd, Gilford and Spring (1966). A few representative values, which may be useful on occasion, are given in Table 13.3, together with data for common building and facing materials.

8.5.1   Unavoidable absorption in studio

The first absorbers to be considered are those which are inevitably present in the studio as a result of its structural design and function. Under this heading come the structure of the room itself, the molecular absorption by the air in the room and the minimal absorption by smooth surface finishes. The absorption of the structure by dissipation of sound energy by resonant vibration is effective mainly at low frequencies, although such components as wood-joist floors and suspended ceilings may have wideband absorption characteristics. One must also take into account absorption by carpets, furniture, curtains etc. if these have to be introduced for other than acoustic reasons, and it must be remembered that persons using the studio, and particularly audiences, if present in appreciable numbers, absorb significantly at middle and high frequencies.

When there are only a few people present, seated or scattered about the studio, values of absorption cross-section per capita may be used, as shown by Burd, Gilford and Spring (1966). For densely seated audiences, however, such figures tend to be inaccurate; in the design of concert halls, Beranek (1962, p. 547) showed that it was more accurate to use average values of absorption coefficients multiplied by the plan areas of the audience, and he found that the values of these coefficients were unaffected even if there were appreciable numbers of empty seats scattered among the seating.

His recommended coefficients are given in Table 8.2.

Later, Kosten (1965–66) showed that very accurate predictions can be made for the whole absorption in a concert hall if the audience area is multiplied by a coefficient that takes into account not only the audience itself or the seats, but also an allowance for the additional surfaces of the hall that are provided for keeping the rain out. The average coefficients of the wall and ceiling surfaces are so low that it introduces little error to assume that these surfaces

Table 8.2 **Absorption coefficients of occupied and unoccupied seating (after Beranek, 1962)**

Frequency	Occupied seating	Unoccupied seating (cloth-upholstered)
Hz		
67	0·34	0·28
125	0·52	0·44
250	0·68	0·60
500	0·85	0·77
1000	0·97	0·89
2000	0·93	0·82
4000	0·85	0·70

bear a constant ratio in area to the audience area, so that they can be accounted for simply by a correction to the coefficients adopted for the seated area. This has little application to broad-actings studios, in which the seating area for audience is generally a much smaller proportion of the whole surface area, and in which the walls and ceiling generally have higher mean absorption coefficients.

Absorption of sound during passage through the air has an appreciable effect on the reverberation time at high frequencies; the term to be added to the surface absorption is proportional to the volume of the room, and, although negligible below 1kHz, it increases rapidly above this frequency, and may contribute the major part of the high-frequency-sound absorption in a large studio. The best recent work on this subject has been carried out in the UK National Physical Laboratory (Evans and Bazley, 1956).

8.5.2 Membrane and bonded absorbers

Low-frequency absorption can be provided entirely by the roofing-felt membrane and bonded-panel absorbers described in Section 8.2. The membrane type, with perforated hardboard covers, is necessary for the lowest frequencies, and the bonded 100–200 Hz.

8.5.3 Wideband porous absorbers

These absorbers consist of a layer of dense rockwool or similar material over an airspace typically 15cm deep. The airspace is partitioned by divisions of hardboard, as shown in Fig. 8.13, into a series of equal cells, so that the particle movements in the air-space are confined to directions substantially perpendicular to the plane of the rockwool. This has the effect of extending the low-frequency absorption (Ingård and Bolt, 1951) by preventing transverse airflow in the space. However, it also prevents the air from traversing the porous material in directions far removed from the normal, and therefore reduces high-frequency absorption.

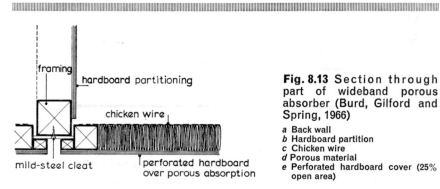

Fig. 8.13 Section through part of wideband porous absorber (Burd, Gilford and Spring, 1966)

a Back wall
b Hardboard partition
c Chicken wire
d Porous material
e Perforated hardboard cover (25% open area)

The spacing of the hardboard separators is therefore designed to arrive at the most suitable compromise between extreme low- and high-frequency performances. Absorption-coefficient tables for various spacings are given in Table 13.3, and Fig. 8.14 shows graphs of typical absorption characteristics.

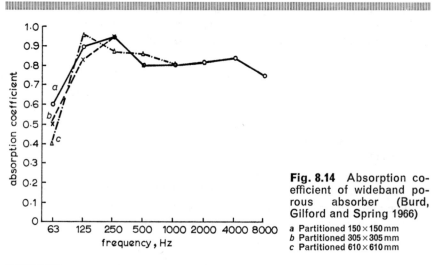

Fig. 8.14 Absorption co-efficient of wideband porous absorber (Burd, Gilford and Spring 1966)

a Partitioned 150 × 150 mm
b Partitioned 305 × 305 mm
c Partitioned 610 × 610 mm

8.5.4 Porous absorbers with perforated or fabric covers

Absorbers of this class are used for the main absorption of middle- and high-frequency sound. The absorption characteristics may be determined by variation of four parameters: the thickness and the flow resistance of the porous material, the depth of the airspace behind it and the characteristics of the cover. The cover may be of a hard material with circular perforations or slots, or it may be a fabric.

An airspace covered with a perforated board and containing a porous material behaves as a highly damped bank of Helmholtz resonators, each hole being associated with the appropriate portion of the airspace behind it. If the area of the holes represents a fraction p of the total area of the cover, and the holes are regularly distributed over the surface (usually in a rectangular array), the frequency of resonance (eqn. 8.1) may clearly be written as

$$(c/2\pi)\sqrt{(p/l'd)} \quad . \quad . \quad . \quad . \quad . \quad (8.17)$$

where d is the depth of the air space and l' is the thickness of the cover plus the end correction.

A suitable variety of absorption characteristics may be obtained

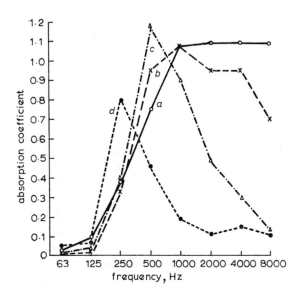

Fig. 8.15 Absorption co-efficient of high-density rockwool, (180 kg/m³), thickness 25 mm over 25 mm airspace (Burd, Gilford and Spring, 1966)

a No cover
b 25% hardboard cover
c 5% hardboard cover
d 0·5% hardboard cover

by the use of boards in which p is 0·005, 0·05 and 0·25, respectively, together with a dense rockwool as the porous material, with a density of about 150 kg/m³ and a flow resistance of 50 SI units. Fig. 8.15 shows a family of absorption characteristics for a series of such absorbers with the range of perforations quoted above. It will be seen from the Figure that all the perforated covers impose a high frequency cutoff, although it is not serious for the highest percentage perforation. However, it is generally found that the reverberation time for a studio rises to an unacceptable extent at high frequencies if the high-frequency absorption is wholly of this type, and it is generally necessary to provide also an area of fabric-covered absorbers or drapes to avoid this fault.

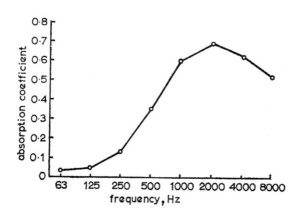

Fig. 8.16 Absorption co-efficient of typical carpet (woolcord)

It is often advisable to cover the floors of studios with carpet for purely operational reasons. This provides an appreciable amount of middle- and upper-frequency absorption, but the absorption is usually highest in the middle frequencies (Fig. 8.16). The combination of absorbers must be carefully chosen to make sure that this does not lead to a serious reduction in reverberation time in this region. By building up laminar materials in such a way that Helmholtz and membrane resonances occur above and below this peak, respectively, an absorber can be produced that is complementary in characteristic to that of carpet over a wide range, so that a very level reverberation characteristic is obtained (Gilford and Druce, 1959). Such special absorbers are, however, rather more expensive than the simpler types dealt with above.

8.5.5   Use of ceiling voids for low-frequency absorption

Low-frequency resonant absorbers are now often installed above transparent false ceilings that enclose voids accommodating ducting and other services. The low-frequency absorbers are mounted on the soffit of the structural ceiling, while the false ceiling consists of perforated plates or tiles, part of which can be overlaid with a blanket of rockwool to provide absorption at middle and high frequencies. Fig. 8.17 shows the absorption

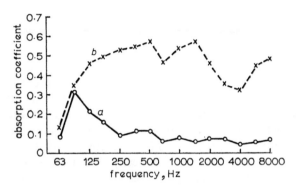

Fig. 8.17 Absorption co-efficient of composite ceiling

One-third of structural ceiling covered by membrane absorbers with false ceiling of perforated plaster tiles, half area overlaid with rockwool blanket
a Membrane absorbers only
b Complete ceiling

characteristics of such a ceiling, curve *a* being measured before completion of the false ceiling and curve *b* referring to the completed structure. The low-frequency absorbers cover about one-third of the soffit, and about one-half of the false ceiling is overlaid with rockwool. The mean absorption characteristic can be adjusted by altering the proportion of the false ceiling that carries the rockwool.

Chapter 9

Diffusion: distribution of absorbers

9.1 Introduction

It has previously been noted that the sound field in a room is generally far from statistically uniform. We have seen that decay curves are usually characterised by fluctuations and that the placing of absorption samples in reverberation rooms must be governed by certain rules if the results from different laboratories are to agree. We have seen, also, that a perfectly regular unfluctuating decay curve, far from indicating isotropic conditions in a room, shows the existence of a single overriding room mode in which all particle velocities are in the same direction. We now examine departures from the statistically random state, to assign a measure to the state of randomness and to find out how it is affected by the absorbing treatment and other features of the design of the room.

'The state of diffusion' is the term used to describe the randomness of the sound field in a room. The field is said to be perfectly diffuse if it has uniform distribution of sound energy throughout its volume and if the directions of propagation at arbitrarily chosen points are wholly random.

If the sound field in the room takes the form of a single standing-wave pattern, it fulfils the first condition of perfect diffusion since the sum of the potential and kinetic energies of any elementary volume in the field will be constant at all points. The second condition will clearly not be satisfied, because all the particle velocities are in the same direction.

Any improvement in the state of diffusion of a room will therefore imply the breaking up of strong standing-wave systems, the diversion of energy from strongly excited modes into weaker ones and the reduction of points of high intensity caused by focusing sound into particular places or directions.

The definition given above describes only a perfect state of diffusion. It is clear that it gives no assistance in assigning a numerical value to a state that is less than perfect; this is to be expected, because the imperfection may be associated with poor distribution either of intensity or of direction. It would be as difficult to assign a number to describe the condition as to derive a quantitative measure of ill health, whether the ill health takes the form of an infectious malady, an accident or a fatal organic disease.

Here we are in a little difficulty, because, without a reliable form of test, it is impossible, logically, to ascribe any particular effect to the state of diffusion, and, without a clear idea of what to look for, it is not possible to establish an objective test. This is probably the reason why it has taken half a century of acoustics research to approach the stage where diffusion can be both measured and correlated with subjective effects. It is first necessary to establish correlations between observable properties of the sound field and to show that one or more of the more highly correlated of them reaches a maximum or minimum value for the conditions described in the definition of the perfectly diffuse field.

It is generally accepted that two studios that are similar in their reverberation characteristics may be different in their sound. One may have a less acceptable sound than the other, and it is found that it can be improved, the sound being made less coloured, less 'bathroomy' or more even from point to point, by adding to the walls and ceiling scattering shapes, such as coffering, hemispheres or hemicylindrical pilasters. These additions need not be absorbers, and may not change the reverberation time by an amount corresponding to the change in the acoustics. It is reasonable to assume that the initial differences between the studios are due to different states of diffusion, since differences in absorption or reverberation time have been eliminated. The same changes can be produced by changes in the distribution of absorbers, or by the modification of surfaces that cause focusing.

Many attempts have been made to quantify statistical properties of a room of a kind that one would expect to be affected by the state of diffusion. Examples are the depth or frequency of fluctuations in the steady-state-transmission characteristics between two points in the room, the depth of the fluctuations in decay curves, the variation of intensity or reverberation time from point to point in the field etc. Most of these prove to be mainly dependent on the reverberation time and even independent of the diffusion. For instance, the number of fluctuations per cycle bandwidth in the steady-state-transmission characteristic is related only to the reverberation time and the distance between the points between which the characteristic is measured (Section 7.5.5). Generally speaking, these methods have not been particularly successful, largely because there was not a prior reason for assuming any definite relationship between the characteristic under investigation and the state of diffusion.

9.2   Measurement of diffusion in large enclosures

Three methods of measurement having direct validity in relation to diffusion in large enclosures have been established. Meyer and Thiele (1956) measured the intensity of sound reaching a highly directional microphone as a function of direction relative to the direction of the source. From these measurements, they constructed 'Hedgehog' models of the sound field by pegging rods of lengths proportional to the intensity into holes in a sphere, aligned with the directions of measurement.

If A_n is the length of a rod and the total number is N, the mean length is

$$\sum_1^N A_n/N = M$$

and the mean deviation about the mean length is

$$\Delta m^2 = \sum_1^N |A_n - M|/N \quad . \quad . \quad . \quad . \quad (9.1)$$

Putting $\Delta m/M = m$, Meyer and Thiele defined $d = 1 - (m/m_0)$ as the 'directional diffusivity', where m_0 is the value of m in a free-field room. They used d as an indication of the confusion of the original sound by reflections from other directions. Note that the diffusivity is greatest when the lengths of the rods are substantially uniform regardless of direction.

Similarly, Furduev and Ch'eng T'ung (1960) devised a method of measurement, in which a directional microphone is rotated in a steady random noise field maintained by a loudspeaker in the studio under investigation. The electrical output of the microphone is plotted against the azimuth angle on a polar-chart recorder, the direction of the source being taken as the zero of azimuth. The measurement is then repeated in a free-field room, the input to the chart recorder being adjusted so that the reading at zero is equal to that previously obtained in the studio. Fig. 9.1 shows the type of chart that is obtained; curve *a* is that from the studio, and curve *b* is from the free-field room and is thus the directional characteristic of the microphone.

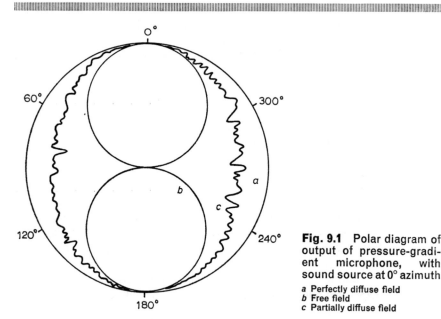

Fig. 9.1 Polar diagram of output of pressure-gradient microphone, with sound source at 0° azimuth

a Perfectly diffuse field
b Free field
c Partially diffuse field

If the studio were perfectly diffuse, curve *a* would, by definition, be a circle, similar to curve *c*. For all other states between perfect diffusion and zero reverberation, the polar diagram from the studio will lie between curves *b* and *c*. The index *d* of diffusion is defined as

$$d = (A_1 - A)/A_1 \quad \cdot \quad \cdot \quad \cdot \quad \cdot \quad (9.2)$$

where A_1 is the area between the circle and curve *b*, and *A* is the area between the circle and curve *a*.

If curve *a* is a circle, $d = 1$, for an axial mode $d = 0$. An attractive feature of Furduev and Ch'eng T'ung's index is that it is amenable to theoretical analysis. They were able to deduce some interesting results concerning the indexes obtained from microphones with figure-of-eight and cardioid polar diagrams.

The use of the correlation function of two microphones at different points in a sound field has been suggested in the past by several authors, such as Gershman (1951) and Gilford and Greenway (1956). The crosscorrelation coefficient between the outputs of two microphones in a free field into which a band of random noise is being radiated decreases as the difference of the distances of the microphones from the source increases. In a reverberant but diffuse field, the correlation coefficient falls with the distance between the microphones, becoming negligible outside distance that we may call the correlation distance. Cook *et al.* (1955) showed that, for a field that is isotropic in three dimensions, the cross-correlation coefficient is given by

$$\bar{R} = \sin kr / kr . \quad . \quad . \quad . \quad . \quad (9.3)$$

where k is the mean wave number of the band of frequencies and r is the distance between the microphones.

For a field that is completely random in two dimensions,

$$\bar{R} = J_0(kr) \quad . \quad . \quad . \quad . \quad . \quad (9.4)$$

where J_0 is the Bessel function of zero order and the first kind.

If the field is not completely diffuse, and contains a standing-wave system, for example, points further apart than the correlation distance will show abnormally high crosscorrelations.

Dämmig (1957) investigated these relations theoretically and experimentally as a measure of diffusion, but concluded that the

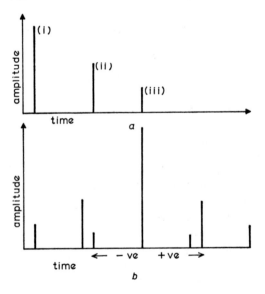

Fig. 9.2 Kuttruff's (1963) method for obtaining correlograms

a Impulse response of dead room with two reflectors
  (i) direct sound
  (ii) ⎱ reflections
  (iii) ⎰
b Autocorrelation display obtained by replaying reversed recording of *a* from original loudspeaker position

correlation coefficient is not entirely dependent on the state of diffusion and that the method is insensitive to small changes in diffusion. In small studios, it may be impossible to separate the microphones by more than the correlation distance, unless a wide band of noise is used. If, however, the band is wide enough to excite several standing-wave systems, the correlation coefficients of these components will be equally positive and negative, so that the increase over the coefficient to be expected with a diffuse field will be small. This restricts the utility of the test for small studios.

Kuttruff (1963) proposed a simple method of obtaining auto- or crosscorrelation functions, without the need for accurate variable delays. A short impulse $\delta(t)$ of pressure is radiated into the studio by means of a loudspeaker, and the impulse response of the room $h(t)$ at the position of a microphone is recorded on magnetic tape. The tape is then played backwards through amplifiers into the loudspeaker, and the time-function $V(t)$ received by the microphone is again recorded.

Then

$$V(t) = \int_{-\infty}^{+\infty} h(-t)h(\tau-t)\, dt \quad . \quad . \quad . \quad (9.5)$$

$$= -\int_{-\infty}^{+\infty} h(t)h(\tau+t)\, dt \quad . \quad . \quad . \quad (9.6)$$

$$= \int_{-\infty}^{+\infty} h(t)h(\tau+t)\, dt \quad . \quad . \quad . \quad (9.7)$$

which is the autocorrelation function of $h(t)$, the impulse response of the room. This may be displayed on an oscilloscope and photographed. It takes the form of a time function that is symmetrical about zero time. Peaks on this display, which will occur at equal positive and negative times, correspond to standing-wave patterns or other regularities in the field, and, therefore, by their amplitude, give a measure of the departure from perfect diffusion. Fig. 9.2 shows the display produced in a room that adds two reflections to the direct sound.

9.3   Effect of distribution on efficiency of absorbers

9.3.1   Introduction

The measurements described above are designed to put a quantitative value to the state of diffusion in a room based on a strict interpretation of the definition of that property. The first two methods are concerned only with the directional aspect of the property, and the third with the presence of standing-wave systems, which also affect the directional distribution of particle velocity in the studio. They are all more applicable to large studios than to small studios.

A continuing problem in the design of small studios is the variation in the absorbing cross-section of a sample of material placed in different rooms or in different positions in the same room. These variations are greatest with small rooms and cause serious errors in design. Moreover, small rooms are characterised by colourations due to isolated strong modes, and it is necessary to understand how these are influenced by measures to improve diffusion and how to assess independently the effect of such measures.

9.3.2 Variation of absorption coefficient of reverberation-room samples

The variation of apparent absorption coefficient has been touched on in Chapter 8 in connection with standardisation of reverberation-room measurements of absorption coefficient. Fig. 9.3, curve *a*, shows the measured absorption coefficient of a sample of rockwool, $10\,\text{m}^2$ in area, placed on the floor of an otherwise empty reverberation room. Curve *b* shows the result of a measurement taken on the same material after hanging in the room a number of sheets of hardboard, randomly oriented, with a

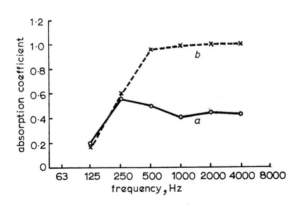

Fig. 9.3 Absorption coefficient of mineral-wool sample, as measured in reverberation room

a Single sample, area = $10\,\text{m}^2$
b As above, but with $34\,\text{m}^2$ of hardboard diffusing sheets hung in room

total area of 34m² (Plate 9.1). The hardboard sheets contributed negligibly to the total absorption, and their only effect could have been to alter the state of diffusion in the room, so that standing-wave systems with particle velocities parallel to the absorbing surface do not escape absorption. Fig. 9.4 shows the results of a similar pair of measurements, except that the sample was divided into four equal areas that were arranged on three mutually per-pendicular room surfaces. It will be seen that the absorbing cross-section of the sample is now greatly increased, with the result that the addition of diffusers has very little effect. The reason for this is clearly that the distribution of the absorber over several surfaces prevents the establishment of strongly excited modes not coupled to the absorber.

Meyer and Kuttruff (1958), using scale models, showed that the effective absorption of a material fixed to the walls of an enclosure can be used as a measure of the state of diffusion.

Randall and Ward (1960) carried out systematic experiments on the mutual effects of patches of absorbent material and of rectan-gular or short hemicylindrical projections on the walls and floor

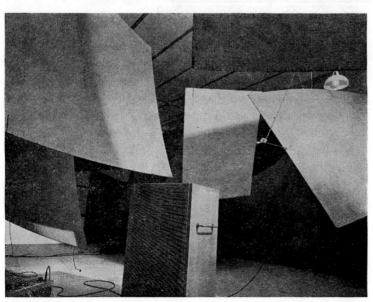

Plate 9.1 Hardboard-sheet diffusers hanging in reverberation chamber

of a reverberation room. They found that patches of absorbent material were as effective in promoting good absorption as perturbations of the walls. Consequently, two patches of absorber on two mutually perpendicular walls of a room were very much more effective than the sum of the two measured singly, unless the field had already been made diffuse by wall perturbations. They accordingly proposed and evaluated a method of measurement of the state of diffusion by comparing the slopes of the top and bottom halves of decay curves.

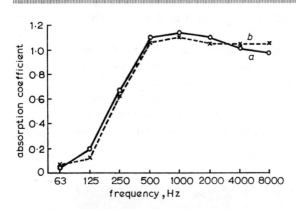

Fig. 9.4 Absorption coefficient of sample, as for Fig. 9.3, but divided into four patches distributed on four walls

a Unmodified reverberation room
b With 34 m² of diffusing sheets

Somerville and Ward (1951) found, also by the use of models, that a projection on a wall affected the diffusion in an enclosure at frequencies down to that at which the wavelength was about seven times the depth of the perturbations. This illustrates the limitations of diffusers of that type, especially for use in talks studios in which good diffusion, down to about 80 Hz, is required for the reduction of colourations. A patch of material causes scattering by reason of the diffraction of sound in its neighbourhood; this was demonstrated in Section 8.3.2 for a Helmholtz resonator, which scatters sound that it does not absorb, and the principle may be extended to other types of low-frequency absorber having dimensions comparable with the wavelength of the sound. Thus patches of absorber of suitable range of action will scatter effectively even if they occupy little depth in front of the studio surfaces. In contrast, a wall perturbation must have a depth of a least 0·6 m to be effective at 80 Hz, and this will be highly inconvenient in a small studio.

9.4   Measurement of diffusion in small rooms

9.4.1   Basis of test evaluation

From Section 9.3, it can be seen that a great deal can be predicted about the state of diffusion in a small room from the arrangement of absorbing materials and wall perturbations. Good diffusion is obtained by having the absorbers distributed in small patches over the three mutually perpendicular pairs of walls and interspersed, if possible, with wall perturbations acting purely as scattering surfaces. Thus it is possible to place a limited number of well defined room arrangements in rank order of diffusion, varying from a very nondiffuse room, in which the absorbing material is on one wall only, to a room that has been made as diffuse as possible by the prescription given above. The conditions can then be tested by any method under investigation, and the rank orders of the results compared with each other and the predicted order of the states of diffusion in the different conditions.

9.4.2   Assessment of tests for diffusion in small rooms

The procedure briefly outlined above was carried out by Ward and Randall (1960). They tried several methods of test, including

(a) the standard deviation of decay rate for different positions of the test microphone

(b) the standard deviation of decay rate for small alterations of frequency

(c) the variation of particle velocity with change of direction (they made use of two different types of directional microphone)

(d) the change in gradient between the early and later parts of decay curves

(e) the variations in measured absorption of an absorbing material in different parts of the room

(f) the mean amplitude of fluctuations in decay curves.

Of all these tests, they found the most promising to be the change of the gradient of a decay curve, which served as an indication of the degree to which different modes of the space were capable of independent decay at their own different rates. The mean rate of decay of energy, averaged over all modes, determines the gradient, and this mean rate decreases as the faster-decaying modes disappear. Fig. 9.5 shows the decay curve for two modes of the same frequency, simultaneously excited, having decay rates in the ratio of 4:1. The mode with the longer decay time is assumed to have an initial level of excitation 30dB below that of the short-decay made. It will be seen that, notwithstanding its lower initial excitation, the long-decay mode dominates the later part of the decay curve.

The measurements made by these authors all depended on the laborious analysis of decay curves or reverberation measurements in great enough numbers to achieve statistical significance.

By the use of the digital processing equipment described in Section 7, Spring and Randall (1969) were able to extend the feasible amount of statistical data and thus reach more decisive conclusions. They equipped a rectangular room so that the acoustic condition could be rapidly altered by changing the positions of approximately 100 low-, middle- and high-frequency absorbing units. They adopted a series of six different conditions in which the whole of each surface of the room was made either reflecting or absorbing. They varied from one in which all surfaces were reflecting to one with all surfaces absorbing, and included conditions in which all absorbers were on either one or two parallel pairs of walls. These two latter conditions were expected to give the worst diffusion. No scattering wall perturbations were used.

The following measurements were made in each condition:

(i) *Reverberation time*

This was automatically derived by the processing of decay curves,

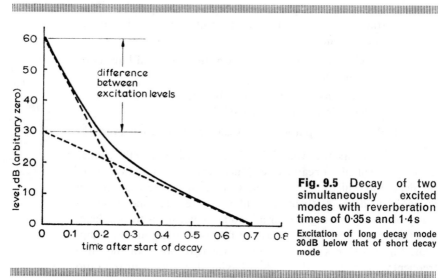

Fig. 9.5 Decay of two simultaneously excited modes with reverberation times of 0·35s and 1·4s

Excitation of long decay mode 30dB below that of short decay mode

and computer analysis (Chapter 7). It was necessary to know the reverberation time accurately, because, as noted above, some apparently promising parameters have been found to be more dependent on reverberation time than on diffusion as such.

(ii) *Standard deviation of reverberation time with position (P_c per cent)*

This was derived by computer during the analysis of ten decay curves at each frequency (for ten microphone positions in the room) and expressed as a percentage of the mean reverberation time for each frequency.

(iii) *Variation of decay rate during the decay process (S_A)*

This was obtained from the digitised decay curves by computing the area between the decay curve and a straight line connecting the starting point of the curve to the point at which it had fallen to a level 6 dB above mean noise level. This area was divided by the whole area enclosed below the straight line and above the line 6 dB above noise. This index is shown in Fig. 9.6.

(iv) *Furduev and Ch'eng T'ung's index (d)*

This was derived for microphones with both figure-of-eight and

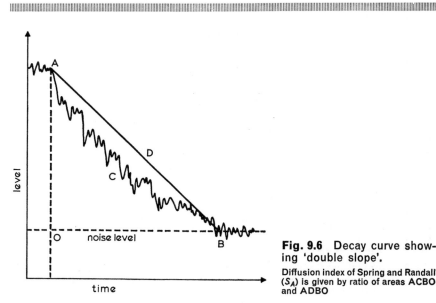

Fig. 9.6 Decay curve showing 'double slope'.

Diffusion index of Spring and Randall (S_A) is given by ratio of areas ACBO and ADBO

G*

cardioid polar diagrams. Measurements were made with octave bands of noise centred at 250 and 500 Hz, and at $\frac{1}{3}$ octave bands of noise centred at 1, 2 and 4 kHz.

Other parameters were derived, but did not produce consistent enough results to warrant further consideration.

Fig. 9.7 shows the six experimental conditions of the room and summarises the values of the three parameters $P_c \%$, S_A and d. In conditions $a-e$, there was carpet all over the floor with a mean absorption coefficient of about 0·5 over the range 250–4000 Hz; in condition f, there was no carpet, a plywood floor being exposed.

Of the parameters investigated, the index d showed a strong correlation with reverberation time, while P_c and S_A were not so correlated. The rank orders of the six conditions are shown beside the values of P_c and S_A, and agree to within one place, except for

room condition	reverberation time at 500 Hz	value and rank order		
		P_C	S_A	d
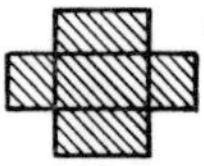 a	s 0·32	14 % 3rd	0·93 2nd	0·37 —
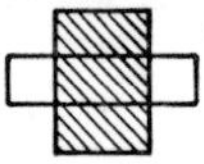 b	0·44	35 % 6th	0.85 5th	0·39 —
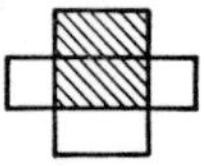 c	0·52	16 % 4th	0·80 6th	0·45 —
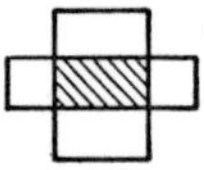 d	0·56	24 % 5th	0·87 4th	0·48 —
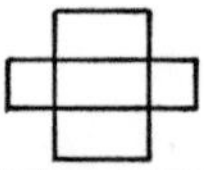 e	0·82	8 % 2nd	0·89 3rd	0·51 —
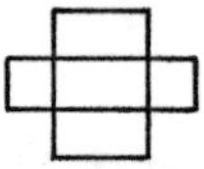 f	1·50	7 % 1st	0·96 1st	—

Fig. 9.7 6-room configurations used by Spring and Randall (1969). Diagrams in 1st column show walls and ceiling; floor was carpeted except for condition f.

condition *c*. They are in agreement in so far as they place the empty room first, and one of the conditions having reflecting walls in one direction only as the worst. The authors conclude, therefore, that these two parameters are the best available indexes of diffusion in small rooms. Both were easily calculated by computer analysis from digitised decay curves as obtained in the course of reverberation measurements, and therefore required no additional studio time in the course of routine tests.

9.5 Effect of position and size of absorber on its performance

9.5.1 Effect of position

The sound field in a room may be considered at any instant as composed of a series of standing waves corresponding to the natural modes of the room, excited to various degrees by the sound from the sources in the room. All these modes will have antinodes of pressure at the walls of the room, except where the modal particle velocity happens to be parallel to one or more such walls. Since the various modal velocities add in random phase, the mean sound pressure will be greater near the walls than in the middle of the room, and still greater at the edges and in the corners. Waterhouse (1955) has shown that the sound-pressure level at a wall is 3 dB higher than in the body of the room, 6 dB higher at the edges and 9 dB higher in the corners. Wöhle (1956) has confirmed these figures experimentally, and also (Wöhle, 1959) examined the behaviour of a Helmholtz absorber with a small neck placed in these special positions.

It was shown in Chapter 8 that the maximum absorbing cross-section of a Helmholtz absorber with a small orifice is

$$\rho c / R_r$$

where R_r is the radiation resistance of the resonator in the sound field. Now the power absorbed at resonance is proportional to the product of the square of the pressure and the absorbing cross-section, and the pressure at an edge is $\sqrt{2}$ times that in the middle of a wall. On the other hand, Wöhle (1959) showed that the radiation resistance in an edge is doubled compared with that at the centre of a wall. The maximum absorption at resonance is therefore unchanged. Since the radiation resistance is doubled, however, the bandwidth of the resonator, when matched to the room (Section

8.3.2), is also doubled; hence the effective absorption by the resonator, considered over a sufficient bandwidth, is doubled at an edge and similarly doubled again at a corner. To be strictly accurate, the peak absorption does increase somewhat in the edge and corner positions, and it occurs at successively lower frequencies in these positions, so that there is rather greater advantage than that to be expected simply from the increase of the bandwidth.

If the test sound consists of a band of noise, the excess pressure vanishes rapidly at distances from the wall, edge or corner exceeding a quarter of the wavelength of the midband sound. Waterhouse shows that, in a cubical room with a side of length 2·5 times the mean wavelength, only about 6% of the total surface can be considered as being 'near' to an edge in this manner, and 1% 'near' to a corner. Therefore, it is only low-frequency resonant absorbers of the Helmholtz or, to a lesser extent, panel or membrane type, which can be made to work at appreciably higher efficiency by mounting them in edge or corner positions. Very little area of porous absorbers, operating chiefly at higher frequencies, and therefore at shorter wavelengths, can be accommodated in these special positions.

9.5.2   Effect of size

It was noted in Section 9.3 that the division of an area of absorber into several patches on mutually perpendicular surfaces yields higher effective absorption coefficients, unless other means of improving the diffusion of the sound field are present. A comparison between Figs. 9.3 and 9.4, however, shows that the subdivided sample has, at 700 Hz, a peak of absorption considerably higher than that reached by the single large sample even in conditions of good diffusion. This excess absorption is attributable to the effect of diffraction at the edges of the sample. The sample is softer than the surrounding wall surfaces, and therefore the lines of flow in the immediate neighbourhood tend to converge to the sample, just as water in a bath converges towards the plug hole. The sample therefore receives incident sound from a larger cross-section of the field than its actual area. This would be evident from the derivation of the absorption cross-section of small low-frequency absorbers in Chapter 8.

Fig. 9.8 is redrawn from Kuhl (1960). He measured the random-incidence absorption coefficient of a sample of rockwool in several successive states of subdivision. As the individual areas become smaller, so does the peak absorption increase, eventually reaching a maximum value 70% higher than that of an infinite-area sample.

Kuhl showed that this effect increases as the impedance of the material decreases, and that the excess absorption due to the subdivision is a steadily rising function of the ratio of the perimeter of the sample to its area.

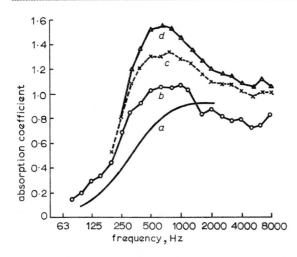

Fig. 9.8 Effect of subdivision on effective-absorption coefficient of 10 m² sample of mineral wool (after Kuhl, 1960)

a Infinite-area coefficient calculated from impedance-tube measurements
b Single sample; edge length = 14 m
c Subdivided sample; edge length = 30 m
d Further subdivided; edge length = 60 m

The subdivision of an absorber therefore has two important effects: it introduces diffusion, and is effective in this respect at all frequencies at which absorption takes place; and it increases the absorption of a given area of material, particularly in the lower middle frequencies, where common types of shallow porous absorbers are inefficient.

9.6 Summary

The effect of an area of absorbing material is thus seen to depend on its size and on its position in the studio, as well as on the arrangement of other materials in the room. Materials are most efficiently used if they are in the form of patches of comparatively small individual area. They should be distributed on various surfaces so as to face in all directions equally. In such a configuration, the diffusion of the sound field will be greatest.

The measurement of diffusion still presents difficulty, particularly in small rooms, because departure from perfect diffusion may take several forms that cannot be assessed satisfactorily by a single test. The best indexes of diffusion appear to be the standard deviations of the reverberation time measured at points distributed about the studio and the curvature of the decay curves. The consequences of these general findings will be considered in the concluding chapters on studio design.

Chapter 10

Design of studios:
general considerations

10.1 Planning

It is no case of special pleading when the acoustics consultant demands to be brought in right from the start of the design of a studio centre or, for that matter, of any building in which sound insulation or acoustics are likely to be of importance. If the early stages of design are carried out without due consideration of the requirements of these factors, it may be very expensive or even impossible to satisfy them at a later stage. Even the choice of site should not be decided without a thorough noise and vibration survey.

It is nevertheless true to say that, in spite of widespread education of the architectural profession in the functional and environmental requirements of buildings, too many mistakes, which result in serious shortcomings in sound insulation or internal acoustics, are still being made in the basic design of buildings. For those who find this statement beyond the bounds of credibility, Backus (1971) quotes an example, as recent as 1967, of an auditorium, built with an elliptic plan form against the express advice of the acoustic consultant, which had to be so heavily treated with sound absorbers to suppress the inevitable echoes that, on completion, it was quite unsuitable for music. It is always a great temptation to the architect to make appearance the primary criterion in the design and to follow current trends in external style, leaving the structural engineer to look after the safety of the building and the environmental engineer and the acoustic consultant to do what they can when the design or even the construction of the shell is virtually complete. One has only to contemplate the vast areas of glass in modern office buildings up and down the country, behind which the occupants are grilled in summer and frozen in winter, to realise the pre-eminence of contemporary appearance. Traffic noise roars in through the single glazing and the lighting varies from good to unbearable.

If this monograph only serves to remind the reader that acoustics, noise and sound insulation cannot be relegated to the position of an afterthought, at least in the design of studio centres, its publication will have been worth while.

The steps that must be considered in planning a new studio centre, or a studio in an existing building, are as follows:

(a) The noise and vibration due to nearby airports, local traffic, especially on urban motorways, underground or surface rail traffic, local industry etc. must be assessed

(b) The studio buildings should be designed to have adequate protection against these noises, either by provision of effective external sound insulation or, for instance, by the use of other buildings as screens

(c) The studio centres should be planned, as far as possible, to separate the technical and studio areas acoustically from public or service areas, and from offices, plant rooms, stores or workshops capable of generating noise or vibration

(d) Within a studio building, music studios should be separated as far as possible from each other and from other types of studio by less sensitive areas or by corridors. If circulation requirements preclude this, provision should be made between the areas for fairly massive multilayer walls that give up to 70 dB mean sound-level reduction

(e) In addition to the working space for each studio, space must be allowed for floating it against structure-borne noise transmitted from other parts of the building and, where necessary, from traffic vibrations. The inter-area sound insulations must also be considered at this stage and room allowed for suitable partitions and floated constructions. In general, each studio will be separated from its neighbours by a structural wall forming part of the main solid shell of the building, and will have an independent inner wall integral with the floated floor. Space must be allowed within the storey height for a false ceiling containing low-frequency-sound absorbers and services

(*f*) Space must be allowed for quiet ventilation systems serving all technical areas, including room for large-section ducts and attenuators.

It is clear from an examination of these six points that they must all be taken into consideration from the start if difficulties and dissatisfaction are to be avoided.

On the other hand, the acoustics consultant should not assume that his requirements are the only ones that matter. The architect will have to keep in his mind questions of cost, circulation, provision of services, hygiene and appearance. It is rather fashionable to take the standpoint that if a design of a building, a piece of machinery or any other article fulfils its function perfectly, it will on that account look right. This view of functionalism is too naïve, for, although it may appear to apply in certain fields of design, it takes no account of independent aesthetic considerations, such as the use of symbolic forms, compatability with the surroundings, or a natural desire on the part of the architect to produce an individual and recognisable edifice. One can say, however, that this type of functionalism is at least more rationally based than the contrary view that outward form is merely symbolic and unrelated to function. For examples of this type of thinking, the reader is advised to refer to Morris (1971).

It is part of the function of an engineer responsible for one of the aspects of a complex design to go as far as possible in meeting the parallel requirements of the design and in giving the architect the greatest possible freedom to work out his conception of the plan in its functional and aesthetic aspects. The acoustics designer must know how to modify his own plans to meet other requirements where they tend to conflict with his own, but he must also know when and how to say 'thus far and no further' when further compromise will result in unacceptable acoustics or sound insulation.

For this reason, it is necessary to know as accurately as possible not only the optimum design parameters, but also the limits on either side of the optimum to which may be shifted before degradation starts to be noticeable to a sensitive listener or viewer. In this monograph, an attempt has been made to define these limits, so that satisfactory designs may be carried out with the minimum interference to the functions of the building in other respects or with its interior or exterior appearance.

Some optima, such as zero background-noise power or infinite transmission loss between two areas, are unattainable; in these cases, one can only quote the maximum or minimum acceptable values. The maximum permissible background-noise level has been discussed in Chapter 3, and the minimum transmission loss between particular areas in Chapter 4.

Reverberation times in studios have optimum values, which will be given in Chapters 11 and 12; the permissible deviations from these are not given, since methods of use vary from programme to programme and even from year to year or producer to producer. Generally speaking, a uniform change in reverberation time of about 10% over the whole range of frequency is noticeable to a reasonably alert listener, and relative differences of this order between different parts of the frequency scale are easy to detect. However, disregarding cost, the amount and effectiveness of the acoustic treatment in a studio can be altered fairly easily without recourse to any major building operations.

Diffusion (Chapter 9) has no accepted means of numerical specification, because there is yet no agreed method of measurement. The diffusion should be at least adequate to ensure that the acoustic treatment is giving the maximum effective coefficients and that the realised reverberation time will agree with its designed value. There seems little evidence for the view, sometimes expressed, that a lesser degree of diffusion may be preferable for certain types of studio.

10.2 General design procedure

10.2.1 Planning of site

The choice of site is a compromise between a large number of factors, from accessibility for staff, performers and public to the type of subsoil. It is almost inevitable that the site will be noisy, because noise is highly correlated with accessibility. If there is an underground railway in the city, the studio centre will probably be built immediately over a tunnel and adjacent to a station serving several branches of the system. Heavy city traffic will pass the front. Land will be very expensive at a central city site, and the site must therefore be kept small; consequently, it will not be possible to find any part of the site where noise or vibration from the principal sources is much attenuated by distance. A site without flanking roads will, however, be an advantage, because it may be possible to use the frontage for administrative buildings partially screening studios further back. The buildings facing the main road will require double glazing, even so, and a forced-ventilation system to ensure tolerable working conditions.

Ideally, there should be a separate works block containing the heating-system boilers, scenery workshops, stores and other potentially noisy operations. The large studios should be in a building separate from the smaller studios used for talks, news etc. An arrangement such as this will avoid the transmission of sound from loud musical forces to the most sensitive areas, and is also convenient from the standpoint of economical layout and circulation of staff, artistes and public audiences. If the whole centre is to be in a single building, the works portion should be on independent foundations and be separated from the rest by a cavity wall with leaves at least 300 mm thick as a barrier against airborne and structure borne sound and vibration.

10.2.2 Planning of studio blocks

The remarks on site planning did not take account of aircraft noise, because, if this constitutes a probable source of interference, its effect will be the same in all parts of the site. The use of separate blocks for the large and small studios nevertheless simplifies protection against aircraft noise in principle, because the small

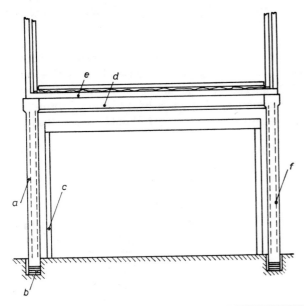

Fig. 10.1 Mounting of one large studio over another

a Columns
b Stable resilient mountings
c Double-leaf wall
d Void
e Base floor of upper studio
f Infill panels

studios can generally be screened from above by less sensitive rooms, while the large studios can be provided with separate roofs with a transmission loss of 70 dB or more (Chapter 4). If studios are mounted above others, it has been found necessary in some cases to support the upper studios independently on columns straddling the lower ones and standing on independent foundations. Such an arrangement has been used in the Munich studio building (Struve, 1964), and in the BBC Television Centre (Fig. 10.1).

It is most economical to arrange the layout of studios and technical areas in such a manner that areas of similar sound-generating potentiality and background-noise requirements are grouped together. By this arrangement, the number of high-insulation party walls is kept to the minimum.

With the general principles in mind, detailed layout of the studio areas can proceed, allowing for walls, floors and ceilings of the appropriate transmission losses (Chapter 4 and Fig. 4.2). All studios and the majority of other technical areas should be floated on resilient layers or mountings to avoid disturbance by structure-borne impact noise. Generally, the resilient layers can be simple blankets of mineral fibre or cork, but those studios standing on ground subject to low-frequency vibration should be floated on springs to give up to 3 mm static deflection (Chapter 5).

The sizes of the studios must be decided in relation to the type of programme, the number of performers, audience etc., and the optimum relations between the dimensions as laid down in Chapters 11 and 12.

The next priority is the design of doors and observation windows. Each entrance to a studio should comprise two doors with a sound lobby between them. The lobby should be acoustically treated with efficient absorbing material that has a broad band of absorption, and the corridors into which they open should have acoustically

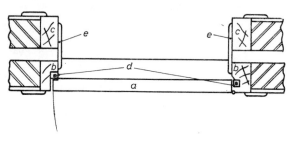

Fig. 10.2 Horizontal section of typical door in double-leaf wall

a Door
b, c Frames
d Magnetic seals
e Cover plates

treated ceilings and soft floor coverings. The design of doors and observation windows should be such that they do not form sound bridges across the wall cavities. A typical design for a door is shown in Fig. 10.2. Space should be allowed for observation windows with double or triple glazing and the reveals should be lined with sound-absorbing material, such as felt. Fig. 10.3 shows a typical design.

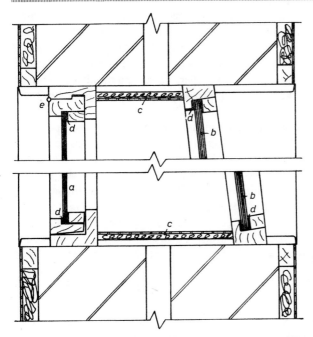

Fig. 10.3 Vertical section of observation window

a Opening light
b Slanting fixed light
c Absorbing reveals
d Compression seals
e Piano hinge

10.2.3 Planning of ventilation systems

The mechanical and thermal design of ventilation systems to give the required air circulation, room temperature and humidity will not be dealt with here, as it is outside the scope of this monograph. The noise characteristics of large systems have been briefly mentioned in Chapter 3, but a few supplementary remarks are required in relation to the broad design of the studio centre. It has been seen that the systems for television studios are principally employed in getting rid of the heat generated by the lighting and that, to do this, very large volumes of air must be moved each second with the minimum of noise generation. This entails provision of space for large-section supply ducts to the upper part of the studio, and a network of extract ducts connected to grilles in the lower parts of the walls. For radio studios, much smaller systems are used, their main function being to provide fresh air for the occupants of the studio. Nevertheless, if low duct velocities are used, in the interests of low turbulence and noise reduction, the space allowance for the ductwork will be just as high in relation to the volume of the studio as in the case of a large television studio. If two studios are close together, care must be taken to prevent transmission of sound between them through the duct-work. Lining the ducts for a distance of about 2 m from the studio inlets or exhaust grilles usually provides sufficient attenuation, provided that there are no substantially straight sections running directly from one studio to the other. To serve a number of studios, it is usual to take the air supplies through separate branches from a main duct to the individual ventilated areas.

Insufficiently lined ducts may also support organ-pipe reso-nances, and thus cause colourations on transmissions from the studios that they serve.

One method that has been suggested for reducing the volume and noise of ventilation systems, but never fully evaluated, is to use small-section ducts with high air velocities. The advantages are that smaller fans, of the axial type, may be used, and, although more sound power will be produced in the ducts as a result of in-creased turbulence, both the fan noise and the noise of the turbulence

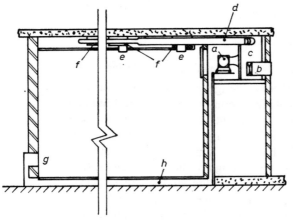

Fig. 10.4 Experimental high-velocity ventilation system

a Axial fan
b Fresh-air inlet and flow control
c Lagged chamber
d Duct
e Outlets
f Expansion boxes
g Return grille
h Return duct

will be of higher frequency than with the low-velocity systems, using centrifugal fans and large-cross-section ducts, and therefore will be very much more easily reduced near to the studio outlets. Moreover, the absence of low-frequency vibration and noise, together with the smaller dimensions, allows the fans to be located nearer to the studios that they serve than is the case with conventional systems. In a small experimental system built to the specification of the author and colleagues in the BBC Research Department, the fan discharged directly into a plenum chamber built into one wall of the studio, and yet the resulting noise level in the studio was below the appropriate criterion curve for talks studios. The velocity in the 130mm diameter ducts was 17m/s; transmission of low-frequency sound was negligible, and high-frequency noise was eliminated by means of a simple labyrinth silencer. Near to the studio inlets, the velocity was reduced to 2m/s by expansion of the duct section (Fig. 10.4).

10.2.4 Planning of acoustics

The final stage of the design concerns the internal acoustics of the studios and other technical areas. The acoustical considerations governing various types of technical area will be outlined in Chapters 11 and 12. When the reverberation characteristic of each studio has been decided on, an estimate must be made of the total absorption by the shell acting as a resonant structure. Some typical structural absorption coefficients are given in Table 13.3 to aid the calculation of this quantity, but a check should be made by reverberation measurements, as soon as the shell is complete. This is necessary, because the variations in the details of the structure are so wide that a really satisfactory estimate is almost impossible.

A word of warning is necessary here. It may seem to be a simple matter to come along one night, as the builders are departing after the day's work, to cover the empty frames for doors and windows with pieces of plywood and settle down to a few hours of purposeful reverberation measurement. This could be all right, perhaps, but let the innocent acoustics engineer accept the kind offer of a passing operative to obscure the window and door holes or to provide temporary lighting, and he will find that he has caused a claim of, perhaps, £1000 to be made by the building contractor for interruption of schedules or for time out. This is no exaggeration. You see, the man who was 'ordered' to fill the holes was unable to finish the job of screeding the floor in the next studio. The screeding therefore had to be finished after the weekend and the carpenters who had been detailed to start battening the walls had to be stood off for 2 days while the floor hardened. So follows the monstrous chain of events that had been started by the first simple offer of assistance. The details differ on every occasion, but the end result differs only in the magnitude of the claim.

It is wise, therefore, to forestall the possibility of such claims by having testing stages written into every contract. This may involve some hard discussion between the architect and the quantity surveyor, but it can usually be done.

The measured absorption of the structure is then corrected for absorption by the surfaces and used instead of the initial estimate. The calculations for the remainder of the absorbing treatment are then adjusted accordingly.

The next occasion for measurement comes when most of the low-frequency absorbers have been added. This measurement will

Plate 10.1 Interior of small studio, showing flush treatment of walls

allow for adjustment to the added bass absorbers. A false ceiling such as that described at the end of Chapter 8, comprising high-frequency absorbers or perforated panels concealing low-frequency absorbing units, allows such adjustments to be made easily and cheaply. If possible, a further measurement should be made when most of the absorbers of all kinds have been installed, but before the final decoration. Any carpets that are to be fitted should be laid temporarily during this measurement.

There are great advantages in using prefabricated absorbers mounted in a timber framework to form a completely flat surface over each wall, using dummy panels of blockboard or similarly solid material to fill in the areas where absorption is not required. The various kinds of absorber are scattered about the walls in the interests of good diffusion and adjustments to the reverberation time and the state of diffusion can be made by replacing or exchanging individual units. This system has been described in more detail by Brown (1964), who has designed many studio treatments of this kind for the BBC.

The framing that supports such a system of flush-mounted absorbers must be rigid enough not to give rise to unwanted low-frequency absorption at its resonances. Such materials as plasterboard battened out from the walls have a similar effect when used for filling spaces between shallow absorbers, and should be used only with due regard to this possibility.

Plate 10.1 is a photograph of a flush studio treatment.

10.3 Use of scale models in planning and design of studios

10.3.1 Principle

The use of scale models for acoustic investigations originated more than half a century ago. Early examples made use of spark photography of wavefronts in ripple tanks and Schlieren photography in air. The development of recording opened up the possibility of using scale models as actual auditoria to add reverberation to a nonreverberant programme.

Suppose that we construct a faithful scaled model of a studio, using the same materials, filled with air and having all dimensions exactly one-nth of the original. The condition for acoustic similarity will be fulfilled if the time scale is reduced in the same ratio as the linear dimensions. This means that all frequencies must be increased in the ratio of $n:1$. If this is done, all wavelengths will be in the same relationship to the model interior, and the reverberation times will be unchanged relative to the altered time scale.

The absorption coefficients of all the materials must have the same values at the scaled-up frequencies as those of the studio materials at the original frequencies. By measurements in a scaled-down reverberation room, it is possible to build up a collection of of scaled absorbers equivalent to those in actual studio use (Brebeck *et al.*, 1967).

The only important property that is not correctly scaled in this manner is the absorption of the air. As will have been seen from Chapter 8, the air-absorption index changes rapidly with frequency, and becomes excessive in relation to the scheme of scaling. To overcome this, it is necessary to dry the air so as to reduce its absorption. A relative humidity of 3–5%, for instance, gives correct air absorption for a model of one-tenth scale.

Objective tests are carried out on such a model at test frequencies in the ratio of n:1 above the normal range. In this manner, it is possible to measure the reverberation characteristic, impulse response, transmission characteristics etc. and, in fact, make all the normal acoustic measurements. The first use of a scale model for subjective listening tests was by Spandöck (1934), who used a gramophone with variable speed to effect the necessary frequency transformation. The method is to record the test programme, which may, for example, be music or speech recorded in a studio with a low reverberation time, and play this into the model at n times the original speed, rerecording the output of a microphone in the model at the same elevated speed. The record is finally reproduced for listening at the original speed of recording.

Improvements on his technique have mainly resulted from the development of plastics-tape magnetic recorders and from improvements in transducers and electronic equipment as a whole. Attention to the details of the models, particularly in the provision of accurately simulated absorbers, has also contributed. The technical difficulties are formidable: for a 1/10th-scale model, one must have loudspeakers and microphones capable of faithful reproduction from about 400 Hz to 100 kHz, and the recorder must perform satisfactorily over this range using two different tape speeds. A relative humidity of about 3% must be maintained in an enclosure, which may be as large as 100 m³ in volume. To ensure the validity of the model, it must be a perfect representation in every detail, the characteristics of all materials and structures being faithfully scaled or simulated.

10.3.2   Recent work on acoustic modelling

The most intensive work has been carried out by a team at the Technische Hochschule at Munich, led by Spandöck (1965). Most of the major technical problems were solved to the stage where differences in the acoustic treatment in a model could be heard by the method described in Section 10.3.1. The loudspeakers were electrostatic, of 25 mm diameter. They gave a usable characteristic from 400 Hz to 100 kHz. Up to 90 of them were used to produce a high enough sound level and to simulate an extended source such as an orchestra. It must be said, of course, that a large number of loudspeakers, all fed coherently with the same programme, do not really represent an orchestra in which every individual instrument produces different sounds, differently phased, from the others. In principle, a better representation of an orchestra could be achieved by means of a sufficiently large number of separate channels, and such an arrangement must be the eventual goal. It is already feasible for small musical combinations as, for instance, quartets.

The microphones were of the condenser type, with diaphragms only 6·5 mm in diameter, and therefore of limited sensitivity. Since

Plate 10.2   Interior of Maida Vale studio 1

the self noise of a condenser microphone and its preamplifier is, to a first approximation, independent of the diameter of the diaphragm, and the capacity change for a given deflection is inversely proportional to the diaphragm area, it is difficult to achieve a satisfactory signal/noise ratio from so small a microphone. This factor placed a limit on the usefulness of Spandöck's system; although reasonable differences in the acoustics could be heard, the impairment due to noise was serious enough to make an absolute judgment of the acoustics difficult.

Improved results have recently been obtained by the acoustics team of the BBC Research Department (Harwood, 1970). They used low-noise field-effect-transistor preamplifiers developed by one of their number, K. F. L. Landsowne, which gave a reduction of 15dB in noise level compared with the preamplifier used by previous workers. They chose a scale factor of one-eighth, as against one-tenth used by the Munich team. The larger scale factor had the disadvantage of yielding an appreciably larger model, but it was found convenient in the early stages of development, because it enabled standard magnetic-tape recorders with speeds in steps of 2:1 to be used without modification. The slightly larger scale also eased some of the technical problems in design and operation, such as those connected with the loudspeakers and

Plate 10.3 Interior of scale model of studio of Plate 10.2

microphones. The loudspeakers developed by the Munich team were combined with middle-frequency direct-radiator loudspeakers and moving-coil loudspeakers of 110mm diameter to cover frequencies of 400–3000 Hz. The overall weighted signal/noise ratio of the system was 52dB.

Plates 10.2 and 10.3 show photographs of the largest orchestral studio in the BBC's Maida Vale centre and of a ⅛th-scale model. In preparing the model of the studio, the absorption coefficient of every material making up the inner surface of the studio was measured, or estimated, where direct measurement was not feasible. Materials having similar absorption characteristics at scaled frequencies were then developed, using measurements in a scaled BSI reverberation room as a guide to development. It was found necessary to make the walls of the reverberation room from 12mm sheet steel, to avoid unwanted absorption by resonant vibrations that occurred if less rigid materials were used. The studio model was then treated over its entire surface with the substitute absorbers and panel materials. The air was dried by circulation through an artificial zeolite molecular-sieve material. The use of this material reduced the time taken to dry out the model from a matter of weeks to a half day on the first occasion, and from 1 day to 15min after temporarily opening the model. The reverberation characteristic thus obtained is compared with that of the actual studio in Fig. 10.5, with due shift of the frequency scale.

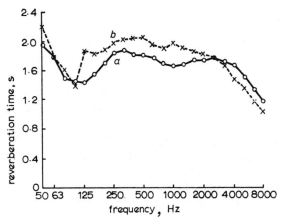

Fig. 10.5 Comparison between full-size and scale model acoustics

a Reverberation-time characteristic of large orchestral studio (Maida Vale 1)

b Characteristic of one-eight-scale model, frequency scale divided by 8

H

10.3.3 Subjective tests using models

For subjective listening tests, it is necessary to start with suitable programme material recorded in very nonreverberant surroundings, so that the only perceptible acoustic atmosphere on the final tape is that added by the model. If one is concerned with large auditoria or concert studios, programmes recorded in a large well damped television studio with a substantially level reverberation characteristics below 1s will provide a reasonable approximation to 'dead' programme material. Recordings of a symphony orchestra playing in a canvas marquee at the Llangollen Eisteddfodd have been found good enough in this respect to serve for the development of artificial-reverberation equipment, for which nonreverberant programme material was essential. Geluk in Holland has made recordings of an orchestra playing in the open air, but the difficulty experienced by players in hearing the rest of the orchestra led to noticeable lack of cohesion in the resulting recordings.

In 1969, by the combined initiative of the Building Research Station and the BBC Research Department, financially supported also by the universities of Bradford and Salford, recordings were made of the English Chamber Orchestra playing in the large free-field room at the Building Research Station. The players were provided with 'foldback' from the programme microphones by way of earphones, though some of the players found that, with practice, they could dispense with this aid to ensemble playing. The recordings were copied on to discs, which are available for purchase. The recordings are described in a paper by Burd (1969).

10.4 Summary

The planning of a studio or studio centre must take in the requirements for acoustics and background noise from the very start. The site and its layout must be chosen to reduce the effect of traffic and aircraft noise on all the studios and the leakage of sound from one studio to another. Noise-producing areas should, if possible, be housed in a block separated from the studio blocks.

The next stage is to decide on the dimensions of the studios and the volumes required for sound insulation, ventilation ducting, sound lobbies and observation windows.

Next, it is necessary to consider the design of walls and mountings for studios against noise and vibration, to achieve specified background-noise levels, and estimates must be made of the structural sound absorption which will be introduced by the shell.

Lastly, one must design the internal finishes with due reference to the requirements for absorption and diffusion.

Provision must be made in the specification and contracts for reverberation and sound-insulation measurements to be carried out at various stages in the construction, and for adjustment of absorption and diffusion prior to the final decoration.

Chapter 11

Design of studios for radio broadcasting

11.1 General requirements of radio studios

Radio broadcasting now consists almost entirely of prerecorded programmes, whether of music, drama or talks. Live programmes, i.e. those broadcast in real time, are confined mainly to news bulletins and 'outside broadcasts' of sports or public events; even these are often heavily laced with recorded inserts.

Classical-music programmes are usually recorded very much as they are played, with little editing apart from the insertion of introductory remarks or chat about the music. Even these verbal annotations are preferred if they are recorded as part of the programme from the studio itself, since only in this way can the correct atmosphere be maintained. Didactic speech programmes or readings from literature are similarly recorded for later listening solely as a matter of convenience.

On the other hand, there are many types of programme which are built, either by assembling extracts from many times their length of recorded material, or by combining several recordings or recorded tracks into one integrated recording. The former type of programme is represented by the documentary and the latter by a majority of the output of popular music.

Multitrack recordings of popular music, which are favoured even more by gramophone-record producers than by their opposite numbers in broadcasting, make use of a number of microphones placed so that each receives, as far as possible, the sound from only one instrument or group of instruments. The tracks are afterwards combined with different amounts of amplification, added echoes, reverberation or various frequency distortions, to produce a desired effect. As a compromise, the tracks can be mixed by an experienced studio engineer at the control desk during the recording.

It follows from this review of the methods by which programmes are brought to their final form that the commonest type of studio in a broadcasting centre is a small talks studio with a listening room equipped for playing back recordings produced elsewhere and for performing editing or dubbing operations. Such studios and their listening rooms are roughly of the dimensions of a fair-sized living room; they serve for the origination of speech programmes of all types, for listening, monitoring, editing and dubbing, and to provide speech inserts or narration for drama and music programmes.

Music studios must be provided in several widely different sizes to suit the widely varying numbers of performers to be accommodated and the type of music performed. The largest, for symphonic and choral music, must be comparable in size with a concert hall.

A studio for drama must be large enough to allow the performers to move in relation to the microphone, thus producing auditory perspective. Specialised rooms must be provided having differing acoustic qualities, and space must be found for effects apparatus.

Thus, a studio centre will comprise a large number of small studios for speech and editing operations, supplemented by special-purpose studios for programmes for which small studios are not suitable. There can be little sharing of studios between differing types of programme, and thus design tends towards increasing specialisation. This situation differs markedly from that in a television-studio centre, as will be seen in Chapter 12.

The background noise and sound-insulation requirements for sound studios of all kinds were considered in Chapter 3, and therefore Chapter 11 will be confined to a consideration of the internal acoustical requirements.

11.2 Design of studios for particular programme types

11.2.1 Studios for speech

With such a high proportion of the programme origination and associated technical operations being carried out in small rooms, it is imperative, on economic grounds, that these rooms should not be unnecessarily large; yet, if they are made too small, the speech quality will be badly impaired by colourations which become less noticeable as the dimensions are increased. A great deal of work has been done to improve the acoustics of small studios, so that their dimensions can be kept small without undue impairment of performance.

The general acoustic properties of small rectangular rooms were described in Section 7.5 of Chapter 7. They were considered more fully by the author (Gilford, 1959) in relation to the design of small talks studios and listening rooms. The conclusions may be summarised as follows:

(a) Rooms of which the dimensions are comparable with the wavelength of sound at speech frequencies are characterised by single-frequency colourations that occur if single prominent axial modes or groups of modes are separated from their nearest neighbours by intervals of 20 Hz or more.

(b) Colourations will be more noticeable if the room lacks adequate diffusion.

These conclusions lead to the following recommendations regarding the dimensions and shapes of small studios:

(a) The dimensions of the studio should be planned to avoid the bunching of axial modes at particular frequencies, with large intervals separating them from neighbouring modes. This may be done by a method of trial in which two convenient dimensions are arbitrarily fixed and the third dimension is chosen to give interlacing frequencies. The dimensions thus fixed are taken as the mean dimensions of the room. The volume should not be less than about $60 \, m^3$, nor need it be above $80 \, m^3$ for acceptable quality, though the OIRT recommends $120 \, m^3$.

(b) Good diffusion is ensured by the suitable distribution of absorbers between the room surfaces and over the individual surfaces. It should be arranged that, over a wide range of frequency, the mean coefficients of absorption of the three pairs of parallel surfaces should be equal to within a factor of 1·4:1. If one pair of facing walls has a relatively low mean absorption coefficient, there is a tendency for long decays and colourations to be heard corresponding to modes in which the particle velocities are perpendicular to that pair of walls, irrespectively of the proximity of modal frequencies associated with other directions. If one pair of facing walls has a

much higher coefficient than the other two pairs, middle-
or high-frequency rings are often heard because of axial
or tangential modes formed by the other two pairs.

On individual surfaces, the absorbers should be arranged in a
series of patches, rather than in one large area. As shown in
Chapter 9, the diffraction at the edges of the absorbers causes
scattering of the reflected sound and, hence, improved diffusion.

An important effect, noticeable in a studio of any size in which
one pair of surfaces is more highly reflecting than the other two
pairs, is the flutter echo. This effect is caused by the repeated
reflection of sounds alternately from the two facing surfaces. It
may often be cured by adding a few patches of absorbing material
on the offending walls. It is seldom or never observed if the mean
absorption coefficients are equal within the limits specified above.
A feature of the flush-wall treatment described in the previous
chapter is that, if flutters are present when a new studio is first
tested, it is a simple matter to alter the distribution of the absorbing

Plate 11.1   Small speech studio, showing patchwise distribution of absorbers
A, B and C are absorbers of different frequency ranges

units over and between walls, and, in the author's experience, a cure can usually be achieved within an hour or so.

Many authors recommend slanting the walls to avoid parallel surfaces. This does not remove colourations; it only makes them more difficult to predict. For the reduction of flutters however, it is effective, and Nimura and Shibyama (1957) have shown that angling of opposite walls to the extent of about 5° tends to reduce the bunching of the room modes at low frequencies. The subjective significance is difficult to discover, and the nonrectangular construction is generally more expensive.

There appears to be no advantage in modifying the walls by adding hemispheres, hemicylinders or other projections to give increased scattering if the patchwise distribution of absorbers as recommended above has been carried out. The action of any diffusing device must extend throughout the colouration region (roughly 80–300 Hz) to be of any practical value. Somerville and Ward (1951) have suggested one-seventh of a wavelength as the minimum effective dimension of a diffusing projection, and this implies a depth of 0·6 m at 80 Hz. Such a depth would be very wasteful of space, and would necessitate larger room dimensions to avoid restricting operational space. Plate 11.1 is a photograph of a typical talks studio, showing the patchwise distribution of absorbers.

The optimum reverberation time for a talks studio is 0·3–0·4 s, depending on its volume, the lower figure referring to the very small studios and the higher figure to a large talks studio of about 150 m³. A reverberation time of 0·35 s and a volume of 100 m³

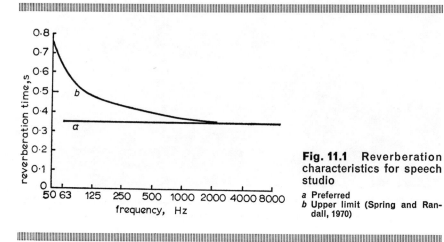

Fig. 11.1 Reverberation characteristics for speech studio

a Preferred
b Upper limit (Spring and Randall, 1970)

are the figures favoured by the author. The time should be nearly independent of the frequency in the range 62–8000 Hz, though a slight rise at the lower end is acceptable if the absorbers are well distributed. A careful experimental study has recently been carried out by Spring and Randall (1970), who found that there are grounds for allowing the reverberation time to rise from 0·35 s at middle and high frequencies to about 0·47 s at 25 Hz and 0·75 s at 63 Hz. This is shown in Fig. 11.1. Curve *a* is the preferred curve and curve *b* is the permissible maximum.

It is a common device to include in the microphone circuits a filter which gives slight bass attenuation, typically 2–3 dB per octave from 250 Hz downwards. This does not remove low-frequency colourations that are frequency-selective narrowband effects. It does, however, restore the balance of the bass end compared with the middle and upper frequencies, which is upset by the normal tendency to listen to loudspeaker speech at a materially higher level than that produced by speech between persons in the same room. This difference, amounting to about 8–10 dB on the average, appears to be a compensation for the absence of the attention or concentration which is stimulated by the physical presence of the speaker. Somerville and Brownless (1949) gave 71 dB as the preferred listening level for reproduced speech, but normal conversational levels are about 60–65 dB(A).

In a typical design of talks studio, low-frequency absorption will be provided, to a minor extent, by the structure and by resonant absorbers of the panel, membrane or Helmholtz type. Helmholtz absorbers are favoured in Germany, but in Britain membrane absorbers have been adopted almost universally, because of their greater predictability. Recently, the BBC has turned towards the use of deep porous absorbers with partitioned airspaces behind and perforated covers with as little as 0·5% open area. These absorbers, as described in Chapter 8, behave as modified Helmholtz absorbers, and give high absorption over a small bandwidth at low frequencies.

In the UK, an appreciable part of the middle- and high-frequency absorption in small studios is provided by carpet laid in most of them for purely operational reasons. European studios are not normally carpeted, and necessarily carry a larger relative area of special absorbers for this range.

11.2.2   Studios for drama

A studio for radio drama should permit the creation of an acoustic 'scene' of any kind, whether indoors or outdoors. It must

H*

enable the producer to create effects of stationary or moving perspective, to produce the aural effects associated with a person talking while moving towards or away from the hearer or to simulate any other acoustic impression which may enhance the dramatic quality of the performance.

In European organisations, the problem is solved by building drama complexes consisting of a number of rooms radiating out from a central control room. Each room has an acoustic treatment corresponding to the main acoustic situations to be represented, e.g. with very short or long reverberation times. A typical drama complex may consist of a 'main' studio, a talks studio for narration, a dead room and a reverberant room. A recording room, a control cubicle and an apparatus room serve for the technical operations. Within these rooms, a wide variety of acoustic scenes can be obtained by making use of microphones with different polar diagrams and by varying the distances of the performers from the microphone.

A more flexible arrangement is adopted in this country by the

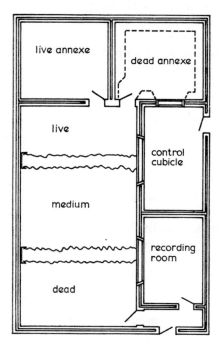

Fig. 11.2 Diagram of drama studio with divided main area and two annexes

BBC. The basic unit is a long studio, the length of which is divided into two or three sections by heavy curtains drawn across from side to side. The sections have different reverberation times, and may be used either separately or together by drawing back the curtains. In addition, there may be a very dead room leading off the main studio and a similarly placed reverberant room, the sections of the main studio being intermediate between these two extremes. Fig. 11.2 is a plan drawing of a studio possessing these features, and Fig. 11.3 illustrates suitable reverberation characteristics for the divisions of the main studio and for the separate rooms.

The dividing curtains are made of heavy, tightly woven, sailcloth material; two such curtains are used together, separated by a gap of about 1m. The acoustic separation which is provided by the curtains between the sections is not very great, normally amounting to no more than 7–10dB at 500Hz, but this is sufficient to ensure that the acoustics of each section are broadly characteristic of that section. As a refinement, the curtains facing the deader end of the main studio may be faced with thick velour to increase their high-frequency absorption, and those facing the more reverberant end may similarly be faced with a smooth impermeable plastics material.

The acoustic absorbers in the sections of the main studio may be of any of the types used in other studios. The same principles of distribution hold as elsewhere, and particular care must be taken to avoid flutter echoes, which will be very damaging to the acoustic

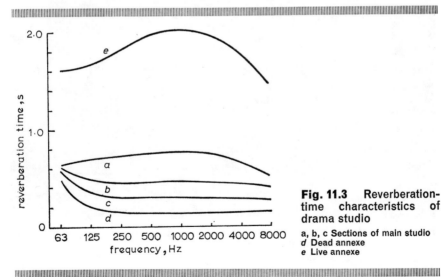

Fig. 11.3 Reverberation-time characteristics of drama studio

a, b, c Sections of main studio
d Dead annexe
e Live annexe

environment. An effort should be made to achieve a good balance between the upper- and lower-frequency absorbers in each of the sections, although it is found difficult, in practice, to install a great enough area of low-frequency absorbers in the deadest section of the studio. The curtains generally offer little transmission loss at very low frequencies, and thus the low-frequency reverberation time in the dead section tends to be increased by sound diffusing back from beyond the curtains. This is one good reason for the provision of a separate, closed, dead studio, which usually opens off the live end. The aim in treating such a studio will be to attain the lowest possible reverberation time at all frequencies, and the treatment will consist of thick layers of highly absorbent fibrous material or of porous wedges of up to a length of 0·5 m.

The separate reverberant room will be untreated except for a small area of absorbers designed to operate at quite low frequencies, at which absorption by the surfaces of the wall and floor materials are negligible.

One of the most difficult acoustic effects to simulate in a closed

Plate 11.2 Interior of drama studio, showing dividing curtains, observation windows and flush wall treatment

studio is, not surprisingly, that of the open air. This is often ideal-ised as a total absence of reflections, and is simulated by placing the microphone in a highly damped studio with thick carpeting on the floor. However, this does not give a convincing acoustic picture of the open air. In the first place, there are always reflections from the ground; a thinly carpeted or even a bare floor therefore im-proves the realism, and the reflections from the floor help to mask the residual low-frequency reverberation, which cannot be removed without the use of great depths of absorbing treatment. Secondly, open air is not usually free from reflections but only from close reflections, a characteristic which it shares with the interiors of very large halls, such as the Royal Albert Hall, London. The sound of a river boat's siren may be followed by reverberation of up to several seconds in duration, made up from multiple reflections from trees, houses, river banks and various topological features. Lastly, the open air is not usually silent; it carries a multitude of sounds, amounting together to quite high levels, including sounds of distant traffic, the barking of dogs and the singing of birds. In some drama studios in Germany, recordings of such sounds are deliberately added to enhance the realism of open-air scenes. The more prominent of these sounds are accompanied, of course, by the characteristic reverberation.

A unique feature of the drama studio is the presence of all sorts of apparatus to provide incidental noises and effects as required by the play. The details are outside the scope of this monograph, but it is relevant to mention that it may take three main forms. Apparatus in the studio designed to make the desired noises straight into the microphone has been largely supplanted, but remaining examples are the inevitable timber or iron staircases, the gravel walks, doors, windows and water tanks. With the improvement of sound recording and the perfection of instant-start devices for the replaying of disc or tape recordings, it has become simpler to use recordings in place of live effects. These are inserted in two forms: they may be added by electronic means into the programme chain or they may be played by loudspeaker into the studio (acoustic effects). In the latter case, the acoustics of the studio are added to the original recording, which is thus made more compatible with the acoustic atmosphere already established.

11.2.3 Studios for orchestral music

It is necessary to make a distinction between the requirements for the broadcasting of 'classical' music on the one hand and of 'popular' music on the other, even though there is naturally a good deal of overlapping in the classification. 'Classical' music in this context will be taken to mean music composed specifically for performance in the presence of an audience with no idea of reproduction in mind. Such music requires, by common consent, a 'natural' microphone balance, which gives to the listener the impression that he is hearing the music from the body of a good concert hall or music room, sitting at such a distance that the instrumental parts achieve a good blend and that the direct sound is enriched by a suitable admixture of reverberant sound from the hall.

The term 'popular' music will be used to denote all music which is intended for performance in association with electroacoustic devices; it therefore includes not only 'beat' music which depends for its effect on amplification by batteries of loudspeakers consuming kilowatts of electric power, but also that large class of arrangements and special compositions which depend upon the use of several microphones by which any instrument or combination of instruments can be brought into prominence, often in an unnatural manner.

Obviously such a classification is, at best, crude, but it will serve to illustrate the needs of the two extreme types between which fall individual works. It is the task of the studio manager to create the best conditions for each case. All musical works made up to the 1920s played in their original form clearly fall into the 'natural balance' category.

It must be realised that a composer does not work in a social vacuum; he instinctively takes into account the way his composition will be heard. Palestrina did not compose his church music for an open-air bandstand, nor did Verdi expect his operas to be performed in St. Peter's Church in Rome. Truisms such as these could have been deduced by anyone with the slightest musical sensitivity from the nature of the compositions. Even Bach's organ works from the Leipzig period show a significant change from the earlier Weimar period, of a kind one would expect from the more reverberant nature of the Thomaskirche compared with the Weimar churches.

It is therefore not an unexpected result that, even in a monophonic broadcast which is considerably less realistic than a stereophonic version, those listeners who are accustomed to hearing live orchestral music show a distinct preference for a single, relatively distant microphone over arrangements consisting of more than one microphone. Jones (1967a) reached this conclusion by listening tests in which some 200 subjects, all accustomed to concert going, heard and compared recordings made simultaneously in a good studio using different microphone arrangements. A single crossed pair of directional microphones show the same advantage in stereophonic recordings over the most sophisticated multimicrophone arrangements. This need cause no surprise unless we can believe that the classics were composed for the benefit of listeners equipped with nine or ten ears, which could be extended on long stalks so as to sample the sound field in the neighbourhood of the various instruments.

There may be some excuse for juggling with several microphones in a studio which has really serious acoustical defects, especially for monophonic reproduction. A particular hall, for instance, which was built for religious services, and is now very popular for gramophone recordings, gives such an overreverberant and muddled sound that a single distant microphone would be quite unsuitable. Very good results are nevertheless produced by the use of several microphones.

It may be found in a studio that is otherwise satisfactory that there is a deficiency in the tone of one section of the orchestra, and one may attempt to correct this by hanging an extra microphone near to this section. This certainly increases the contribution from the section, but it also upsets the perspective, and may well lead to a chain reaction in which microphones are added one after another in an attempt to improve the ensemble.

One reason often given for the use of combinations of close microphones is that the ears of the musically interested public have become accustomed to the sound of music recorded in this way for commercial gramophone discs. Certainly, an appreciation for this type of sound is very common among those who habitually listen to recorded music and who never hear an orchestra playing in person; this group includes some who write on high-fidelity topics. Their preference is rooted partly in a history of recording with primitive apparatus in boomy studios and partly in the apparent clarity with which the music is invested by the modern application of these methods, in which the orchestral parts favoured at any instant stick out like burnt currants on the top of a fruit cake. Those who prefer to listen to the score rather than to the music probably favour this trend.

The guiding principle for a studio in which classical music is to be performed is that it should be suitable for broadcasts with a single microphone or stereo pair. If a simple arrangement such as this does not produce satisfactory results, it is doubtful if a proliferation of microphones will ever produce the ideal sound. To facilitate good natural balance, the orchestral studio should preferably be of a size comparable with a fairsized concert hall. A volume of 10000 m³ is probably sufficient unless it is desired to make provision for a large studio audience without the acoustics being excessively affected. The reverberation time should be about the same as that of a concert hall of similar size, being longer for a very large studio than for a smaller one. Fig. 11.4, curve *a*, shows the BBC recommendation for the reverberation time of radio music studios (Burd, Gilford and Spring, 1966).

The presence of an audience may be partly compensated for by moving the microphone to a more distant position. It is found that

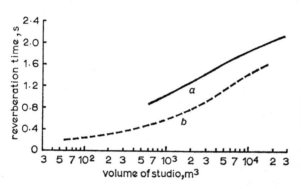

Fig. 11.4 Optimum reverberation time for radio studios; maximum reverberation time between 500 and 2000 Hz (Burd, Gilford and Spring, 1966)
a Music studios for natural balance
b Other studios, e.g. for light entertainment

the ratio of the intensity of the reverberant sound to that of the direct sound (liveness, Chapter 7) should be of the order of 5–8, and it is possible to calculate, by this means, the shift required to compensate for the increased absorption due to the presence of an audience. Hessische Rundfunk's large concert studio at Frankfurt incorporates hinged shutters, the two faces of which have different sound-absorbing characteristics. By turning these in suitable numbers to expose one face or the other, compensation may be made for the audience. This is the only instance known to the author, however, of variable-absorption arrangements which are in regular use.

There is a certain amount of dispute surrounding the reverberation characteristic of an orchestral studio. Somerville and Gilford (1958) maintained that the reverberation time at low frequencies should not exceed that at middle and high frequencies, since excessive bass reverberation induces masking of much of the detail in the middle-frequency parts. Other authors (e.g. Bruel, 1951) favour a moderate rise in the reverberation characteristic below 150 Hz or so, claiming that this adds a desirable warmth to the full orchestral sound. Bass masking can often be avoided, even in the presence of a rising characteristic, by providing absorbing surfaces behind the loud brass and percussion instruments that are usually placed at the back of the orchestra. The absorbers, which may take the form of portable screens, as described by Brown (1965), eliminate reflection from behind, and reduce the effective power of the nearer instruments.

To ensure good definition and warmth of tone, adequate diffusion should be provided. Distribution of absorbing materials in patch formation, as described above for speech studios, may not be sufficient, because only a small area of absorbers is normally required in an orchestral studio. The walls and ceiling should therefore be generously provided with irregularities to scatter the wavefronts during reflection. The most efficient shape in relation to the depths of the irregularities is a rectangular parallelopiped (Somerville and Ward, 1951). Several modern German studios for music at Hanover and Munich are almost entirely lined with rectangular wooden panels over air spaces of several depths. They serve both as low-frequency absorbers and for diffusion (Kuhl and Kath, 1963). Rectangular diffusers in the Maida Vale studio may be seen in Plate 10.2.

On the question of shape, one should be guided by good concert-hall practice. The height of the studio should be rather greater than half the width (Marshall, 1967) and its length should be greater than the width. It should be possible to seat a full symphony orchestra within one half of the studio, hanging the microphone in the other half to obtain perspective and balance. Various claims have been made from time to time that particular exact ratios (e.g. double cube or $1:\sqrt[3]{2}:\sqrt[3]{4}$) are specially favourable, but there is no evidence whatever that exceptionally good auditoria have possessed these dimensions, let alone owed their excellence exclusively to this factor. Certainly, grouping of room modes is unlikely to exercise any recognisable influence, as it does in small talks studios, because, in such large spaces, the modes are so tightly packed together that, at every point on the frequency scale, the bandwidths of several adjacent modes overlap.

A word is necessary on the staging on which the players farthest from the conductor are normally raised. This staging should be strongly enough constructed not to constitute unwanted panel absorbers. On the other hand, staging units carrying violoncello or double-bass players are preferred if they have tops of plywood or other panel material. The action of the plywood is to provide an impedance match between the spike on which the instrument is supported and the air, thus increasing the radiation of sound. Rostra carrying percussion instruments should be made from concrete or other rigid material; any resonance in the rostrum would be heard as a colouration on the tone of the instrument.

11.2.4 Studios for chamber music

Studios intended for the performance of chamber music, including accompanied or unaccompanied solos, should have the same general form as the larger music studios described above. The volume should be very much smaller, partly to preserve the sense of intimacy appropriate to the intended setting of the music and partly because microphone distances need not be so great to cover a small combination. The liveness at the microphone should be similar to that in a larger studio, and it follows that the reverberation time should be lower, roughly in the same ratio as the linear dimensions. The curve of the reverberation time against the volume (Fig. 11.3a) is continuous with that for larger studios.

11.2.5 Studios for popular music

The term 'popular music' covers such a wide variety of types that it is not possible to generalise on the exact requirements of a studio. Moreover, fashions in reproduction have changed cyclically during the last twenty years or so with alternating calls for live and dead sounds, especially for arrangements of light music. The last ten years have seen the introduction on a large scale, of the reverberation plate and other devices (Section 7.5) by which acceptable artificial reverberation may be superimposed on virtually non-reverberant sound. The use of multimicrophone techniques for creating a variety of both monophonic and stereophonic effects has also accelerated a trend towards very much less reverberant studios.

This trend was described very clearly by the studio designer Brown (1965), himself a well known clarinet player, band leader and jazz musician.

'The use of multi-channel recording techniques demands good separation as an *a priori* function in studios . . . Some of this separation can be achieved by the use of cardioid or super-cardioid microphones close to the sources, but it is usual to use twenty to thirty microphones and they cannot all be directed away from unwanted sources. It is therefore necessary to use separation booths or to make the studio very dead and provide adequate acoustic screening between channels . . . The largest orchestras for 'pop' music comprise about 30–35 players and their requirements in physical space, allowing $3\,m^2$ each, are a modest $100\,m^2$. The ceiling need not be high, so that the total volume of the studio need not exceed $300\,m^2$ . . . The economic advantages are enormous and it can be shown that very low reverberation times and adequate separation of the sources ensure that satisfactory results are invariably attainable.'

An objection which has been raised to very dead studios for popular music is that the musicians themselves find them difficult to play in, particularly with respect to keeping strictly in time with the other players in an ensemble. This is undoubtedly true of a musician playing for the first time in such circumstances, and may remain a difficulty in a large television studio, for instance, in which there is a complete absence of audible reflections from overhead. However, playing with other musicians in a small dead studio becomes very much easier with practice, and experienced session players attain the most remarkable precision of timing. It is an advantage if a studio used in this way has only just enough floor area to contain the musicians and their instruments. The reverberation time may be as low as 0·2s at all but the very lowest frequencies. Fig. 11.5 shows a section through a type of wideband absorber

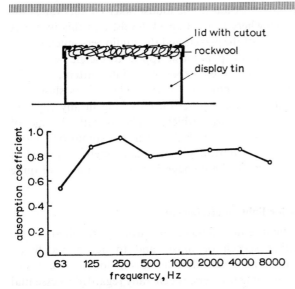

Fig. 11.5 'Biscuit-tin' absorbers for dead popular-music studios

a Section through absorber
b Absorption characteristic

that has been used in the popular-music studios of the BBC and its absorption characteristic. The principle of the absorber is the same as that of the wideband absorbers described in Chapter 8. They are made by stacking together tinplate boxes with skeleton lids, originally intended for the display of biscuits. The fronts are closed with expanded metal backed by dense rockwool, and the walls of the individual boxes partition the whole space behind the assembly, controlling the characteristic at the upper and lower extremes of the frequency range.

To attain such a low reverberation time as 0·2s, it is necessary to use very large areas of efficient wideband absorbers of the type described above, and as much as 80% of the total wall areas may be so treated. In such a dead acoustic environment, it is necessary to make sure that there are no objects in the studio which have long decay times; this may include such things as fire extinguishers, door panels or unattended musical instruments.

A word of warning should be added here before concluding this Section. The sound-pressure levels in the direct sound from the loudspeakers used by many 'pop groups' are far in excess of those which can be considered safe from the point of view of hearing-damage risks. Levels of over 130dB have been measured during peaks, and these are high enough to cause rapid onset of temporary threshold shifts (Flugrath, 1969). Continuous exposure to such levels would inevitably result in permanent loss of hearing sensitivity. The use of compression in the loudspeaker circuits raises the mean level to somewhere near the peak levels, and this increases the effective exposure of the players and the studio personnel. Fortunately, most followers of 'pop' music appear to escape permanent hearing damage because their exposure is intermittent, but the players themselves are continually at risk. These facts should be brought to the notice of every person partaking in any capacity in the broadcasting of music which is played with the aid of amplification giving more than 90dB(A) in the studio or control cubicle. No one should unwillingly or unknowingly be placed at risk in this way, because the damage is both insidious and quite irreparable.

11.2.6 Studios for light entertainment

Studios for light entertainment are usually required to be suitable for either speech or music and to accommodate an audience. The studio audience, although it is undoubtedly a source of irritation to the home listeners, is generally regarded as essential for programmes where the performers depend upon some rapport with, and reaction from, an audience. For this reason, many disused theatres have been brought into use as variety studios, and they fulfil this function admirably. The volumes of these theatres vary between about 3000 and 9000 m³, and the reverberation times of the most acceptable fall close to curve *b* of Fig. 11.4. The equipment will generally include a fairly extensive sound-reinforcement system, because it is found that the best reaction is obtained from the audience if they are submerged in a highly amplified sound field. The installation and adjustment of such systems to get the maximum possible sound levels is a specialised art, which has been treated elsewhere. The acoustics specialist can assist by means of increasing the absorption and by locating and treating areas contributing to instability by the formation of focused reflections in the neighbourhood of the stage.

11.3 Design of other radio-broadcasting areas

11.3.1 Listening rooms and quality-monitoring rooms

Small rooms intended for quality listening as opposed to the origination of speech programmes should resemble the speech studios described above in their general acoustic characteristics. However, they should possess two characteristics which could well be mutually conflicting. First, they must provide the best acoustic conditions for judging and comparing programmes of all kinds and for distinguishing small differences between very similar qualities of programme. Secondly, they should bear a general resemblance, acoustically, to the average listener's viewing and listening room.

It would be a profitable digression here to consider the relationships between monitoring conditions and home listening arrangements. A distinction must be made between the equipment used for the special purpose of listening or viewing and the living room in which the listener is sitting. In deciding upon the type of electro-acoustic equipment to be used for monitoring purposes, the broadcasting organisation must strive to transmit signals of high quality, so that any improvement made by the listener to his own equipment (in certain mutually understood directions such as linearity and frequency range), results in an improvement in his reception. Were the broadcasting organisation, for instance, to monitor programmes by means of an 'average' loudspeaker, improvements in domestic equipment would result in worse reproduction for the listener, and progress would be discouraged. There is a regrettable tendency among television producers to match their sound outputs in this way to the pitifully inadequate arrangements provided in the majority of mass-produced television receivers now in use. The justification of such a policy is usually that a large proportion of television-programme output, particularly of light entertainment, is accompanied by multitrack sound of an artificial nature, and that, without some guidance provided by an 'average' monitoring loudspeaker, the programme will simply not be effective. The principal factor involved is that of 'presence', or the apparent closeness of performer to the listener. This depends greatly upon the high-frequency content of his contribution, and is most upset by the use of a poor loudspeaker. The long-term damage which could be done to programme quality in this way is, however, so serious that the policy should always be to provide the highest-quality equipment for sound monitoring, placing the onus on the individual to make his own arrangements, if he must, for degrading the output.

A rather different philosophy is applicable to listening or monitoring rooms, because the listener's room is normally used for many other purposes besides listening. It would, for instance, be quite unreasonable to monitor all programmes in rooms of which the reverberation time had been reduced to 0·1s, assuming that listeners who wished to improve their reception would have the knowledge and money at their disposal to do the same in every room in which they were likely to listen,

Fortunately, this would, in any case, be inapplicable. In a series

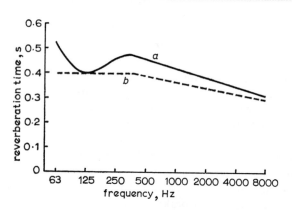

Fig. 11.6 Reverberation characteristics of listening rooms

a Average of 16 domestic listening rooms
b BBC standard curve for control cubicles and quality-monitoring rooms

of experiments by the BBC Research Department (Gilford, 1959), a statistical analysis was made of judgments on five programme sources, using five different listening rooms. It was found that the best discrimination and judgment was obtained if the listening room had a reverberation time of about 0·4 s below 250 Hz, falling steadily thereafter to about 0·3 s at 8000 Hz. This is also typical of a fairly well furnished living room, though the reverberation time of such rooms is rather higher at the bass end for solid-floored rooms and lower for those with wood-joints floors than the 0·4 s indicated by these experiments. Fig. 11.6 curve *a* is the recommended reverberation characteristic, and curve *b* is an average for ten living rooms of various types.

If the listening room is very much more reverberant than shown by curve *a*, the details of the programme will be blurred by the acoustics of the listening room; if it is much lower, the sound will depend too much on the listener's position in relation to the axis of the loudspeaker and on its directional characteristics. Most loudspeakers have greater directionality at high frequencies than at low frequencies, and therefore a dead room tends to make the programme sound 'toppy' to a listener on the axis and deficient in upper frequencies at positions well away from the axis.

11.3.2 Echo rooms

The use of echo rooms in connection with the modification of programme acoustics was briefly mentioned in Section 7.6.3. Such rooms may be used for two basically different purposes: to create unusual or dramatic effects, such as the sound of an underground cavern, or to improve the quality of sound, particularly music, originating in too dead a studio. The first object of use is easily achieved, provided that the reverberation time of the room is long enough, especially since an excessive reverberation time at low frequencies often helps to produce the desired effect. Rooms for the second purpose require more careful design. As mentioned in Section 7.6.3 above, a small reverberation room generally exhibits colourations at low frequencies, owing to the wide spacing of the room modes, and as much as possible must be done to prevent them from impairing the sound of the original programme. A reverberation time of about 1·7–2 s should be the aim. Only a small area of absorbing treatment is normally necessary, and it should be designed to absorb mainly at low frequencies, where separated modes occur. Various shapes and sizes of room have been tried, none of them being found to have clear advantages over the others. The room should be as large as can be allowed on economic grounds, and marginal improvements can be gained by any measures which improve the diffusion of the sound field. Compact sound absorbers will have this effect.

A suitable design of echo room would have a volume of 80 m³ and be rectangular, with dimensions chosen as previously described to avoid prominent isolated axial modes. The walls and ceiling would be finished in tiles or hard gloss paint over smooth concrete or plaster. The floor would be of bare concrete, and the absorbing treatment would consist of two or three membrane low-frequency units, each about 0·5 m square in frontal area and of depths arranged to yield resonance frequencies coinciding with any frequencies at which a peak of reverberation time is shown by measurements on the empty shell. Positions for the loudspeaker and the microphone will be found by trial and error, starting from positions a little way out from diametrically opposite corners. Final adjustments to the reverberation time would be made after the installation of the loudspeaker and microphone, using small individually designed absorption units.

11.4 Studios and listening rooms for stereophony

The acoustical conditions necessary for stereophonic broadcasts or recordings do not differ grsatly from those required for monophonic programmes. Given equal microphone distances in the same studio, a stereophonic programme may sound more reverberant or less than a monophonic reproduction according to the effective polar diagrams of the microphones. A crossed-pair arrangement will generally sound more reverberant than a monophonic or spaced-pair pickup. The design of a studio does not have to be changed, therefore, for stereophonic, as opposed to monophonic, transmissions, either as regards dimensions, reverberation time or the provision of diffusion. The main changes necessary for a stereophonic broadcast will concern microphone placing, which is equally a matter of adjustment from one programme to another of the same type.

There are, however, two main particulars in which a stereophonic studio must receive special treatment. Background noise is specially disturbing on a stereophonic transmission, and noise curves some 3–4 dB lower than those recommended in Chapter 3 for monophonic studios should be the aim. This is particularly important in programmes incorporating a great deal of movement. Secondly, care must be taken to avoid the possibility of strong reflections from walls of the studio near the performers, since these may seriously upset the perceived direction of the sound sources in the studio. Additional wideband absorbers around the studio at performer level will be an advantage.

Similarly, listening and control rooms for stereophony should be very much like those for monophonic programmes described in Section 11.2.2, but should have, in addition, a band of absorbing material about 1m wide round the back and side walls at loudspeaker height. The symmetry of the room about the loudspeakers should also receive attention, both in the design stage and when the room is nearing completion. Moderate room asymmetry does not necessarily affect the symmetry of the stereophonic image, because the image centre is adjusted by means of the balance controls so that only 2nd-order effects remain. However, if there is considerable asymmetry that is effective only, say, at low frequencies, the image may be distorted so that the sound from certain instruments is consistently biased to left or right. The immediate cause may be that one loudspeaker is at a pressure node of a strong standing-wave system. In such cases, the sound field can be plotted along the line joining the loudspeakers for a series of $\frac{1}{3}$ octave bands of noise or pure tones, and a more acoustically symmetrical pair of positions found from the results. In one instance, the author was able to improve the general acoustic symmetry by cutting off the access to a small lobby and by changes in the position of acoustic treatment.

Chapter 12

Design of television studios

12.1 General requirements for television broadcasting

A large television studio presents an extremely complex technical ensemble with interacting requirements for lighting, acoustics, the movements of cameras and microphones, and the organisation of performers and operators. The performers are usually assembled in 'sets' or stages constructed on the floor of the studio and consisting of vertical flats of plywood or other sheet material shaped and painted to resemble the required visual background. One or more television cameras, mounted on electrically driven trolleys (dollies) are deployed near each set, the camera operator riding on the dolly that is driven or pushed by other members of the team. Camera cables trail behind the dolly, or, more rarely, are supported by ropes attached to winches in the ceiling space. The scene is lit by banks of floodlights or individual spotlights, which may be raised or lowered by cables operated by electrically driven winches that are fixed to a metal grid spanning the studio high over the floor.

Instead of stagelike sets, a plain cyclorama or backcloth may be used behind the performers, and outside scenes may be projected by a still or cinematograph projector on to a screen in the background.

Plate 12.1 is a photograph of a large television studio showing sets and lights in position.

The action in the studio is watched on television screens by the producer and heard by him and by a sound mixer, who is usually in a separate room. Both rooms command a direct view of the studio floor through large observation windows.

Television studios may be classified as either 'small' or 'large', the functional difference being between those intended for purely static camera work in a which single camera is trained on to one or a few people, and one which is intended for other types of programme involving moving cameras and several sets or backgrounds. The necessity for camera movements and lighting manipulation demands plenty of space around the working sets and ample height above them. Hence a multiset studio must have a fairly large floor area, say, not less than 25 m × 15 m and a working height of at least 8 m below the ceiling grid carrying lighting and scenery winches.

Plate 12.1   Large television studio

Being of such a large volume and so lavishly equipped, large television studios must be kept as fully used as possible in the interests of economy, and this means that every studio must be, as far as possible, suitable for many types of programme. Segregation of studios for particular purposes is not practicable, though a compromise policy may be effected by making slight variations between studios. The permissible range of variation is not very great, since, generally speaking, all television studios must necessarily be considerably less reverberant than sound studios of comparable size. The reasons for this statement are as follows:

(a)   For the great majority of programmes, the microphones should, for aesthetic reasons, be kept out of the picture. In consequence, they will often be at a greater distance from the performers than would be the case in a radio programme. To restore an acceptable ratio of direct to reverberant sound, therefore, the studio should have a correspondingly lower reverberation time.

(b)   As a consequence of its general-purpose role, the studio must be used to portray all kinds of acoustic 'scenery', and, in particular, it must be capable of simulating the acoustics of a small living room. If the studio were reverberant as well as large, the intrusion of long-delayed reflections from distant parts of the studio, even at low intensity, would completely upset the illusion of intimacy appropriate to the scene. Therefore, the studio must be heavily and uniformly enough treated to suppress such reflections to a satisfactory degree.

(c)   The movements of cameras, lights and scenery, often inseparable from a television production, inevitably create a certain amount of background noise (Chapter 3). Most of these sounds originate at some distance from the programme microphones, but are augmented at the microphone by the studio reverberation.

Generally speaking, these three considerations are satisfied by a reverberation time about one half that which would be suitable for a general-purpose sound studio of the same volume. Fig. 12.1 shows a set of design curves for the determination of the optimum

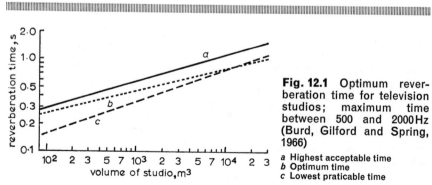

Fig. 12.1 Optimum reverberation time for television studios; maximum time between 500 and 2000 Hz (Burd, Gilford and Spring, 1966)

a Highest acceptable time
b Optimum time
c Lowest praticable time

reverberation time for television studios, plotted against their volume (Burd, Gilford and Spring, 1966). Curve *a* is the highest acceptable reverberation time, averaged over the three frequencies 500, 1000 and 2000Hz. Curve *b* is the optimum with which the studio will be suitable for most kinds of programme that consist mainly of speech. Purely musical programmes present a problem, which will be considered in Section 12.4.

Curve *c* represents, in the authors' opinion, the lowest practicable curve. It is calculated from the following assumptions:

(*a*) The whole floor must be available for the movements of cameras, scenery and people. It will therefore be smooth, hard and nonporous, and consequently a good reflector of sound.

(*b*) Approximately one-quarter of the area of the walls will be required for equipment of various kinds, including electrical outlets, ventilation grilles and controls, observation windows and cable ducts.

(*c*) All remaining areas are treated with a highly efficient wideband absorber with an absorption coefficient approaching unity over most of the audible frequency range.

It will be seen that, for the very large studios, curves *b* and *c* converge. The conclusion from this observation is that it would be virtually impossible to make a satisfactory general-purpose television studio larger than the largest (16000m), at present in use by the BBC. Studios of smaller volume give an increasingly wide latitude in design, for reasons given in Section 12.2.2.

12.2 Acoustic treatment of television studios

12.2.1 Large studios

The general requirements for the treatment of television studios have been discussed in Section 12.1. To achieve a short enough reverberation time for all purposes, it is necessary to use extremely efficient wideband sound absorbers on all available surfaces, and these absorbers must, to satisfy safety officers, have a good resistance to the spread of fire. In most television studios at present, the main roof trusses and most of the structural ceiling are sprayed with asbestos fibre to reduce the conduction of heat to the structure in the event of fire This material provides high absorption coefficients at middle and high frequencies, but there is an increasing awareness of the risk to health associated with the application of asbestos by this method, and it is possible that its use will be prohibited in the near future.

The treatment used on the walls of recent BBC studios are of two kinds, the principles of which have been described in Chapter 8. The main areas consist of wideband absorbers comprising dense rockwool backed by a partitioned airspace. The absorption coefficient of this construction approaches unity, from the highest

Plate 12.2 Area of wideband and bonded bass absorbers in television studio

frequencies that are of interest, down to about 150 Hz. In the bottom octave or so of the required frequency range, where the performance drops off, they are supplemented by bonded bass absorbers consisting of a panel of hardboard 3 mm thick, which is bonded to a sheet of bituminous roofing felt of mass $2.5 kg/m^2$. The composite panel is fixed over an airspace of similar depth (about 180 mm) to that of the wideband absorbers. To comply with fire regulations, the fronts of all the absorbers are covered with a 25 mm layer of starch-bonded rockwool held in place by a steel-wire grid.

Plate 12.2 is a photograph of an area of mixed wideband and bonded bass absorbers. The absorption characteristic of a bonded bass absorber with a rockwool covering is shown in Fig. 12.2; it has a dip at 180 Hz, owing to an antiresonance effect where the sloping portions of the separate curves of the two absorbers intersect. In practice, this dip does not result in an appreciable peak in the reverberation characteristic of the studio, because the loss of absorption at that frequency is small compared with the total absorption in the studio.

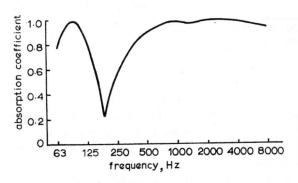

Fig. 12.2 Absorption characteristic of 18cm-deep bonded bass absorbers with 2·5cm of rockwool in front

The absorbers are necessarily disposed in very large continuous areas, to provide as large a total absorption as possible. The method of measuring absorption coefficients recommended in Chapter 8 therefore gives high results compared with the effective coefficients in the studio. The calculation for the prediction of reverberation time may be corrected approximately by using the divided-sample coefficients for all areas within 1m of an edge and reducing the coefficients of the remaining areas by 10%. Fig. 12.3

shows the reverberation characteristic of a large general-purpose studio.

Generally speaking, the presence of scenery and equipment has very little effect on the reverberation time; the additional absorption that they provide is small in comparison with the total built into the studio, but the improved diffusion due to scattering probably has a useful effect.

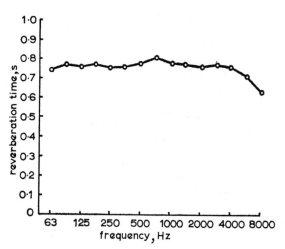

Fig. 12.3 Reverberation characteristic of large television studio

BBC Television Centre, studio 3; volume = 10 000m³

Apart from the desirability of allowing room for the greatest area of absorbers, it is essential to avoid leaving any large areas of untreated wall that may cause echoes. In a large, heavily damped studio, a single reflecting surface, such as a scenery door, may give rise to echoes that are particularly obvious because of the very short reverberation time of the studio as a whole. A cyclorama of hard material is particularly to be avoided, because, being curved, it can cause focused reflections.

The acoustic atmosphere within a television set is determined partly by that of the studio as a whole and also by early reflections from the materials with which the set is surfaced. The usual materials are hardboard, plywood, or canvas on wooden frames, and these materials are painted to give the illusion of 3-dimensional form. The perspective of the sets is manipulated by taking advantage of the fact that the television camera is 'one-eyed', so that a small piece of painted plywood immediately behind the performers may look like a long room stretching out many yards behind them. In these circumstances, of course, a dancer approaching the camera from the back of the set may appear to be equipped with the proverbial seven-league boots. The visual and aural contradictions that result from this sometimes make it difficult to establish a convincing sound perspective; reflections from flats near to the microphone given the aural effect of a small space. Therefore, backgrounds representing larger enclosures should be transparent or absorbing, the necessary long-term reverberation being provided electronically by the methods described in Chapter 7.

A paper by Burd (1968*b*) gives the results of some measurements on common scenery-flat materials. He found that the lighter materials, such as canvas or other fabrics, reflected relatively inefficiently, whereas those having a plywood backing were better reflectors at all frequencies. Differences between various surface finishes affected the reflective properties, mainly at high frequencies. Heavy glass fibre or aluminium were good reflectors.

Outdoor scenes are usually reproduced by back- or front-projection on screens made of fabrics, and these generally present no acoustic problems.

In some television studios, particularly in Germany, an attempt is made, by the use of large areas of Helmholtz absorbers, to reduce the reverberation time at low frequencies to a figure considerably lower than that at middle and high frequencies. The justification of this very expensive operation is the widespread use of directional microphones in these studios. Since such microphones become omnidirectional at frequencies for which the wavelength of the sound is large compared with the dimensions of the microphone, they tend to pick up an excessive amount of reverberant sound at low frequencies, and compensation is made by reducing the reverberation time as described.

12.2.2 Small studios

Small television studios used for static programmes, such as news broadcasts, may receive acoustic treatment similar to that described in Section 12.2.1 for the larger studios. The requirements for very dead acoustics are less stringent, however, because long-delayed reflections cannot occur and the static or near-static operating conditions reduce the incidence of background noise. Concealed microphones can often be used within the field of the camera, and, in any case, for most programmes taking place in small studios, a microphone in shot does not offend accepted conventions.

I

A point to watch when installing concealed microphones, e.g. in a news reader's desk, is that, in such locations, panel or cavity resonances may be heard as colourations on speech. Colourations can also occur as a result of interference between the direct speech and first reflections from surfaces near to the microphone. The use of a directional microphone is not necessarily a solution, because of the deterioration of the polar diagram at low frequencies. The desk should be constructed without any deep or nearly closed cavities, and the top should incorporate a sound-transparent central area designed to carry nothing heavier or larger than a few sheets of script. There should be no reflecting surfaces behind the microphone.

12.2.3   Adaptation of basic studio for different programmes

It will be appreciated from the foregoing that the acoustics and perspective of a scene may be partly created by the intelligent use of set materials. The scope, however, is limited because of the short reverberation time of the studio and the virtual elimination of all long-path first reflections. The studio can be used without alteration for most drama and speech programmes where the acoustics should resemble those of a small room on the one hand or of an outdoor scene on the other. More reverberant conditions must be simulated by electronic or other means. In drama, much use is made of reverberation rooms, reverberation plates or other equipment for adding reverberation when the visual scene suggests a large enclosure with long reverberation.

For the performance of music in a highly damped studio, no means is known of producing an effect that is aesthetically beyond criticism. As discussed in Chapter 11, most music requires a reverberant atmosphere and does not sound satisfactory in dead conditions. In the absence of natural reverberation, all that can be done is to add the artificial kind either in the studio itself or electrically in the programme chain.

However, as shown in Chapter 7, all such devices for adding reverberation to programmes have characteristic imperfections that are particularly obvious when the added reverberation constitutes a large proportion of the total sound energy issuing from the viewer's receiver. This applies to the addition of reverberation in sufficient proportions to create the acoustics of a concert hall on a programme from a highly damped television studio. The only really satisfactory solution is to broadcast orchestral music from a studio or hall with good natural acoustics, but no broadcasting organisation has yet been able to face the cost of a studio reserved exclusively for music programmes. As an alternative, these programmes can be broadcast from a radio studio with suitable acoustics or from a concert hall hired for the occasion. Many such concerts have been broadcast, for example, from the Fairfield Hall in Croydon.

In some musical programmes, it is possible to rely on prerecording the music in a suitable studio and playing it back through high-quality loudspeakers into the studio where any visual action is taking place while the performers mime to the sound of the music. Miming is not entirely acceptable, since the synchronisation achieved between the music and the mimed movements is often obviously imperfect. An alternative method, much used for vocal solos, is to have the soloist and other performers in one studio and the orchestra in another, using closed-circuit television to maintain contact.

One highly important feature of a musical television broadcast is the need to preserve a close match between the visual and the auditory perspectives. The vision producer is always faced with the temptation to present the orchestra and soloists in a succession of changing views and perspectives. The viewer may find himself looking, during successive shots, at a whole orchestra at a distance of 30 m and at the keyboard of a piano from a point directly above it. If the sound of the ensemble does not change between these shots, the contrast between the heard and expected auditory perspectives may be most disturbing.

The worst case is probably that of a closeup picture of a violinist's fingers accompanied by the sound from a distant microphone; in less extreme circumstances, the effect is alleviated by the fact that the field of the eye, in the course of concentrated attention, is only about 1° in diameter, and thus, from a seat in a concert hall, one can concentrate the gaze on a single instrument or scan the whole orchestra at will.

A worse condition is that in which the microphones entirely fail to pick up the sound of an instrument that is in closeup. This is a fairly common occurrence, and cannot be entirely avoided by the proliferation of microphones. A long-focus lens can show a distant instrument with certainty and effectively bring it into the foreground, but a microphone cannot necessarily do so because of the existence of interference. A continual shifting of the sound balance is, however, as detrimental to the entertainment value of the concert as are continual changes in the relative perspectives of sight and sound. The only remedy for these difficulties is artistic restraint in the methods of production.

The subject of music broadcasting on television is a fascinating one, but it cannot be more fully considered here. The basic principles have been described by Alkin, Pottinger and others (1962), who also include material on the historical development of television-studio acoustic treatment.

12.3 Control rooms for television

The usual complement of control rooms for a television studio is three, for sound, vision and lighting. The requirements for acoustics are different in the three rooms.

The vision-control room is occupied by the producer of the programme, secretaries and vision mixers. Their monitoring equipment consists of a number of vision monitors showing the scenes covered at any moment by the several cameras and separate monitors for the picture being transmitted and the picture waiting to be faded or mixed in next. There is also a loudspeaker, which should be of good widerange quality, though the producer is not ultimately responsible for sound quality. The acoustic treatment should be designed to facilitate communication between the producer and the members of his team in the stress of production, and the first object must therefore be to reduce reverberation that will tend to confuse if more than one person is talking at the same time. It is usual to design for a reverberation time of about 0·25 s throughout the range and to reduce the background noise due to ventilation and other equipment to about the levels shown in Fig. 3.1, curve *b*.

The sound-control room should provide good listening conditions for the sound mixer, who is responsible for all matters of microphone selection and placing, for control during the transmission or recording and the use of artificial reverberation or other such devices. He is also responsible for the insertion of recorded effects and for the provision of foldback for performers. The requirement therefore is for listening-room acoustics, as described in Section 11.3, for radio broadcasting.

A special requirement of television, as opposed to radio, control rooms concerns the siting of the monitoring loudspeaker, which must be fairly close to the vision monitors. There is often a window provided between the vision- and sound-control rooms which can be lowered for conferences but which is normally closed to prevent disturbance of the sound mixer by the production business. To avoid obscuring this window and at the same time to keep the loudspeaker clear of obstructions, it may be necessary to hang it in a corner at one side of the window. Although corner mounting is often advocated for loudspeakers, it is not usually practicable to have the effective centre of radiation of a monitoring loudspeaker nearer to the geometrical corner than about 1 m. As a result, interference effects occur between the direct sound from the loud-speaker and reflections from the adjacent walls, causing deep irregularities in the transmission characteristic in the neighbour-hood of the sound mixer's head. Harwood and Gilford (1969) found that this was the cause of complaints of poor quality from monitoring loudspeakers that had given every satisfaction when tested in other situations. The interference can be reduced by acoustic treatment of the reflecting surfaces, but this is not prac-ticable if one of them is the observation window. Alternatively, a cure was effected in particular cases by careful adjustment of the position and angle of the loudspeaker, guided by response/frequency characteristics recorded at the sound mixer's position.

The lighting-control room has no critical requirements for acoustics or background noise.

12.4 Special requirements for sound insulation in television studios

12.4.1 Sound insulation of television control rooms

For operational convenience, the sound-, vision- and lighting-control rooms are usually situated side by side at 1st-floor level along one end of the studio. From this position, they command a good view of the whole floor of the studio. The observation windows between the control room and the studio are of very large area, to give the required angle of view in all directions, and consequently they determine the transmission loss between the two. In the BBC Television Centre, the lower edges of the windows are mounted on an extension to the control-room floor about 0·5 m below the rest of the control-room-floor area and inclined outwards towards the top. This arrangement gives improved visibility of the nearer end of the studio floor.

The transmission loss of the windows and of the extended transoms over them must be adequate to prevent the control-room loudspeaker from being heard in the studio and also to prevent the sound in the studio from being heard directly through the windows and thus misleading the sound mixer. The sound levels used in monitoring are usually below those in the studio for music but above for speech. The interference, if the transmission loss of the window is inadequate, is most likely to affect the control-room sound for music, and the studio for speech.

A mean transmission loss of 44 dB is sufficient to remove such interference for a large television studio where the mutually interfering sources are usually remote from one another, and about 5 dB greater for a small news/interview studio. In the BBC studios mentioned above, an interesting construction was used for the transom extensions over the windows where dry construction of the minimum mass was desirable. It consisted of a number of 1 mm steel sheets interleaved with felt impregnated with a highly damped material on each side of a 100 mm airspace. The mean transmission loss was 45 dB for a total mass of 25 kg/m^2, compared with 230 kg/m^2 for a brick wall giving the same mean sound insulation.

12.4.2 Open-planned recording areas

An essential part of modern television programme preparation is the compilation of news or magazine programmes from filmed or videorecorded extracts of events, interviews etc. This requires the co-ordinated operation of several playback and recording machines while a programme is running or very shortly before the start of the programme. Each machine is provided with its own monitoring loudspeaker for immediate checking of quality and a communication loudspeaker for the receipt of instructions. There are clear advantages in operating the machines together in communicating spaces; two machines are usually situated closely to allow continuity to be maintained during changes of tapes in the longer programmes, and the use of larger numbers in open-planned areas has been tried by several broadcasting concerns. The advantages of flexibility of operation and of easy supervision and co-ordination are, however, outweighed by the interference between the loudspeakers associated with different positions, and this has caused the virtual disappearance of multiple open-planned video-recording and playback areas. In the meantime, various arrangements of screens, sound absorption and directional loudspeakers have been tried in the hope that the sound reaching each operator from machines other than his own will be sufficiently attenuated. The attenuation required to prevent confusion to the operators, according to a brief survey by the author, is about 16 dB. This is the same order as the transmission loss between two adjacent rooms opening into a common corridor with plastered walls. An improvement of about 6 dB may be obtained by covering the walls of the corridor with an efficient sound-absorbing material. Thus, if a number of machines in a large area can be separated by acoustically treated screens about 2·5m in height, so that there are no optical paths between them, a separation of 22dB can be obtained without difficulty, and this is enough for troublefree operation, particularly if the wanted and unwanted sounds are arranged to come to the operator from predominantly different directions.

If the machines are situated so that there is a short optical path between them, no amount of hopeful work with absorbing treatment or directional loudspeakers will produce satisfactory conditions. The minimum practicable distance between an operator and his loudspeakers is of the order of 1m. The sound level at the second operator is assumed to be 16dB lower to avoid disturbance, the distance of the second operator would have to be 6m, even disregarding the effects of reflections. The use of directional loudspeakers would not reduce this enough to compensate for the effects of reverberant sound in the common enclosure. It has therefore been found essential to provide movable partitions to give an extra 10dB transmission loss between machines when operating on different programmes. The final abandonment of open-area working in one organisation (which had adopted this layout against the express advice of its own acoustics specialists) came after a television appearance by a very important person had been transmitted in vision only for the first five minutes because an operator acted on the cue intended for an adjacent machine.

12.5 Future prospects for sound in television

The techniques of television production are advancing from year to year, though the advance is faster in vision than in sound. In radio broadcasting, the only valid output is sound, and there are virtually no arbitrary limits to the changes that can be made in improving the quality and significance of the sound image created by the receiver.

The presence of the picture must always have a restraining influence on the sound. The object of the sound mixer is to produce sound of which the perspective and quality fit the picture so well that the viewer experiences an enhanced sense of reality in the programme. The problems encountered in achieving this in the existing conditions have already been touched on, but they will become more formidable if stereophonic sound is introduced into television programmes, because of the incompatability of the sound-stage width with the angular width of the picture. Thus sound quality and realism are bound to lag behind those available at any time for sound-only transmissions because of constraints or contradictions introduced by the picture.

Radical advances in visual presentation towards improving the compatability of the sound may not be any more likely. For instance, the introduction of stereoscopic pictures combined with large-screen viewing could enable stereophony to be used with success and would also enhance the influence of sound perspective. However, vision in depth would greatly increase the front-to-back distance necessary for sets intended to represent large spaces, and consequently costs of production would rise for this reason alone. The only alternative would be some sophisticated extensions of inlay and overlay techniques using separate backgrounds formed from stereoscopic images. Up to the present, however, a public service of stereoscopic television appears further away than it did 20 years ago, and stereophonic television sound must remain in abeyance.

It is therefore difficult to foresee any major changes in function for television studios in the near future. The tendency, instead, will be towards methods of treatment of studios to reduce their reverberation times coupled with the further development of electroacoustic devices to modify the acoustics or output of these dead studios for programmes requiring a more reverberant atmosphere. There appears to be a steadily increasing interest in programmes about music and musicians (as opposed to set musical performances), and this trend will encourage the attempts to perfect such equipment.

Artificial-reverberation devices, for instance, which at present are all marred by noticeable colourations, will be subject to detailed improvements, and may become more complex. Van Leeuven (1965) has announced improvements on the reverberation plate for the reduction of colouration, and it is still possible that the helical-spring type of device may receive modifications with the same object. Completely new systems hardly seem economic in present circumstances. The author has recently suggested a scheme by which programmes on a continuously moving tape can be speeded up for passing through a scale-model studio with preferred acoustics and slowed down afterwards for replay. To do so, since the whole tape must move at the same speed, it is necessary to divide the programme on the tape into short sections by means of rotating heads and recombine them after processing. The problems inherent in this process are similar to those already solved for video recording systems, and the idea therefore appears feasible. It would nevertheless require considerable development, which is likely to be costly. The success of digital coding and transmission of audio-frequency signals in recent years tempts one to speculate on the possibility of using digital storage and arithmetic for the creation of artificial reverberation that would be capable of infinite variation and adjustment to simulate very good natural acoustics. However, with digital stores obtainable at present, it would be impossible to provide both a frequency range and reverberation time even approaching that which is necessary for acceptable artificial reverberation. At the same time, it must be remembered that the capacity of stores is continually increasing and the cost per stored bit decreasing, so that a device of this kind is not beyond the bounds of possibility. Together with other new principles of operation, such developments could remain an open field of fascinating study in the future.

1*

Appendixes

13.1 Bibliography

This bibliography is classified according to the order in which the subjects are treated in the monograph. Only a few works have been chosen for inclusion in each section, but these should supply all the detailed information that could be desired to supplement the brief outline already given.

13.1.1 Theory of sound

STEPHENS, R. W. B., and BATE, A. E. (1966): 'Acoustics and vibrational physics' (Edward Arnold)

RAYLEIGH, LORD (1945): 'The theory of sound—Vols. 1 & 2' (Dover)

RSCHEVKIN, S. N. (1963): 'A course of lectures on the theory of sound' (Pergamon)

RICHARDSON, E. G. (Ed.) (1962): 'Technical aspects of sound—Vols. 1–3' (Elsevier)

13.1.2 Ear and hearing

BURNS, W. (1972) 'Noise in man' (John Murray, 2nd edn.)

STEVENS, S. S., WARSHOFSKY, F., and the editors of *Life* (1965): 'Sound and hearing' (Time–Life International)

WHITFIELD, I. C. (1967): 'The auditory pathways' (Edward Arnold)

13.1.3 Time and frequency domains, Fourier analysis, correlation etc.

BRACEWELL, R. (1965): 'The Fourier transform and its applications' (McGraw–Hill)

HEAD, J. W. (1964): 'Mathematical techniques in electronics and engineering analysis' (Iliffe)

13.1.4   Acoustic measurements

BERANEK, L. L. (1950): 'Acoustic measurements' (Wiley)
BROCH, J. T. (1969): 'Acoustic noise measurements' (Bruel & Kjaer)

13.1.5   Noise

BERANEK, L. L. (Ed.) (1960): 'Noise reduction' (McGraw–Hill)
HARRIS, C. M. (Ed.) (1957): 'Handbook of noise control' (McGraw–Hill)
KRYTER, K. D. (1970): 'The effects of noise on man' (Academic Press)

13.1.6   Sound insulation

CONSTABLE, J. E. R., and CONSTABLE, K. M. (1949): 'The principles and practice of sound insulation' (Pitman)
BAZLEY, E. N. (1966): 'The airborne sound insulation of partitions' (HMSO)
BERENDT, R. D., and WINZER, G. E. (1964): 'Sound insulation of wall, floor and door constructions'. NBS Monograph 77, US Department of Commerce
PARKIN, P. H., PURKIS, H. J., and SCHOLES, W. E. (1960): 'Field measurements of sound insulation between dwellings' (HMSO)

13.1.7   Structure-borne sound and vibration

CREMER, L., and HECKL, M. (1967): 'Körperschall' (Springer Verlag)
SNOWDON, J. C. (1968): 'Vibration and shock in damped mechanical systems' (Wiley)

13.1.8   Room or studio acoustics

SABINE, W. C. (1922): 'Collected papers on acoustics' (Harvard University Press)
FORD, R. D. (1970): 'Introduction to acoustics' (Elsevier)

13.1.9   Sound absorbers

ZWIKKER, C., and KOSTEN, C. W. (1949): 'Sound absorbing materials' (Elsevier)
BURD, A. N., GILFORD, C. L. S., and SPRING, N. F. (1966): 'Data for the acoustic design of studios', *BBC Eng.*, 64, Nov.
EVANS, E. J., and BAZLEY, E. N. (1964): 'Sound absorbing materials' (HMSO)
'The insulation handbook' (Lomax, Erskine & Co., published annually)

13.1.10   General acoustic design

FURRER, W. (1956): 'Bau- und Raumakustik für Architekten. (Basel–Stuttgart); also English translation of revised edn. (1964): 'Room and building acoustics and noise abatement' (Butterworths)

PARKIN, P. H., and HUMPHREYS, H. R. (1958): 'Acoustics, noise and buildings' (Faber & Faber)

MANKOVSKY, V. S. (1971): 'Acoustics of studios and auditoria' (Focal Press)

13.1.11   Transducers

GAYFORD, M. L. (1970): 'Electroacoustics, microphones, earphones and loudspeakers' (Butterworth)

13.2   Some typical noise sources

This table is divided into two parts, the first giving the sound-power levels of unified sources, generally those inside a studio centre, and the second giving the peak sound-pressure levels due to external or scattered sources where this method of specification is more appropriate. Although only those sources that have fairly characteristic levels and spectra are included, the figures must be taken as representative only; designs should preferably be based on direct measurements in the actual situation.

Table 13.1   Sound power levels in decibels relative to 10^{-12} W, in octave bands

Centre frequency of band, Hz	63	125	250	500	1000	2000	4000	8000
Male speech, voice raised for clear enunciation only	66	73	78	80	77	71	65	65
Male speech, declamatory	$\simeq$ 8 dB higher than above							
String quartet	80	84	88	90	90	88	86	80
Symphony orchestra	96	100	104	109	106	104	102	98
Pop group, amplified	110	125	125	125	120	115	110	100

Table 13.2 **Sound pressure levels, in octave bands, decibels relative to 2×10^{-5} N/m**

Centre frequency of band, Hz	63	125	250	500	1000	2000	4000	8000
Jet airliner flying at 300 m	95	100	100	100	100	97	95	90
Urban motorway, 40 m from edge	82	82	81	78	75	71	71	70
City street traffic, 10 m from kerb	92	92	92	87	86	86	86	84
Scenery-construction workshop, at walls	77	80	83	85	85	85	85	80

13.3 Sound-absorption coefficients

The sound-absorption coefficients of building, facing and furnishing materials that are likely to be encountered in studio construction are tabulated here. A few examples of sound-absorbing materials and constructions designed for adjusting the final characteristics of the studio have also been included. Very much fuller lists will be found in Burd, Gilford and Spring (1966), which also gives brief recommendations for design objectives, and in Evans and Bazley (1960). The coefficients below were measured by the reverberation-room method, using, where applicable, samples divided into three or four areas distributed on several surfaces. For easy reference, the absorbers are separated into eight classes, with the notation used by Burd *et al.*

Table 13.3 **Sound absorption coefficients**

Frequency, Hz	63	125	250	500	1000	2000	4000	8000
(a) Structural absorption, by vibration of whole mass The following figures should be added to surface-absorption coefficients:								
115 mm brickwork, plastered	0·08	0·11	0·05	0·05	—	—	—	—
75 mm clinker block, plastered	0·09	0·13	0·16	0·03	—	—	—	—

Frequency, Hz	63	125	250	500	1000	2000	4000	8000
(*b*) *Common building materials*								
Brickwork	0·02	0·02	0·02	0·03	0·04	0·05	0·07	0·10
Smooth plaster	0·02	0·02	0·02	0·02	0·03	0·03	0·04	0·04
Softwood, painted	0·05	0·06	0·07	0·09	0·10	0·10	0·12	0·15
Soft fibreboard, lightly decorated	0·04	0·06	0·10	0·15	0·19	0·21	0·21	0·22
3 mm hardboard on 25 mm battens	0·03	0·32	0·43	0·12	0·07	0·07	0·11	0·18
(*c*) *Air absorption*								
Values of 4 m (per metre) at 50% relative humidity:	—	—	—	—	0·003	0·006	0·023	0·092
(*d*) *Audience, furniture etc.* (absorbing cross-section, m²)								
Each person*	0·15	0·35	0·40	0·45	0·45	0·45	0·45	0·45
Upholstered seats	0·12	0·24	0·27	0·28	0·32	0·37	0·39	0·39
Plywood seats	0·00	0·02	0·02	0·02	0·04	0·04	0·03	—
(*e*) *Carpets and curtains*								
Woollen carpet	0·03	0·06	0·20	0·45	0·60	0·70	0·65	0·60
Velour curtains, draped to approximately half area	0·05	0·06	0·31	0·44	0·80	0·75	0·65	0·60
Lightweight curtains, slightly draped	0·01	0·05	0·10	0·20	0·50	0·70	0·65	0·60
(*f*) *Low-frequency resonant absorbers*								
Roofing-felt membrane absorbers:								
Single layer	(see Fig. 8.8)							
Double layer (150 mm deep)†	0·91	0·60	0·53	0·35	0·15	0·15	0·15	0·15
Bonded absorbers	(see Fig. 8.9)							
(*g*) *Acoustic tiles*								

(*g*) *Acoustic tiles*

The following figures are representative of types of proprietary tile normally intended for suppression of noise and excessive reverberation:

19 mm cane-fibre tile, perforated surface	0·10	0·20	0·60	0·70	0·75	0·75	0·75	0·75

* For large audiences, see Section 8.5.1

† 0·97 at 90 Hz and 0·52 at 180 Hz

Frequency, Hz	63	125	250	500	1000	2000	4000	8000
30mm glass fibre, with perforated metal front	0·10	0·25	0·50	0·75	0·85	0·80	0·80	0·75
19mm 'fissured' tile, plaster and glasswool	0·10	0·35	0·50	0·70	0·80	0·85	0·90	0·90
12mm expanded polystyrene‡	0·05	0·05	0·15	0·40	0·35	0·20	0·20	0·20
19mm glasswool with fine porous facing 180mm airspace	0·50	0·75	0·85	0·90	0·80	0·80	0·75	0·60

(h) *Porous materials, unfaced*
Rockwool, high-density (145–160 kg/m²):

	63	125	250	500	1000	2000	4000	8000
25mm thick, mounted on wall	0·05	0·10	0·26	0·69	1·00	1·00	1·00	1·00
50mm thick, mounted on wall	0·18	0·34	0·83	0·97	1·00	1·00	1·00	1·00
50mm thick, over 180mm airspace	0·27	0·68	0·88	0·87	0·81	0·83	0·85	0·75
25mm polyester foam	0·05	0·15	0·40	0·65	1·00	1·00	1·00	1·00
25mm polyester foam, faced with thin poly-vinylchloride film	0·05	0·10	0·45	0·65	0·40	0·45	0·35	0·30
25mm sprayed asbestos	0·10	0·25	0·45	0·60	0·70	0·85	0·85	0·85
12mm vermiculite plaster	—	0·08	0·15	0·30	0·45	0·60	0·65	0·65

(i) *Porous materials with perforated facings*
There are so many combinations of materials, thicknesses spacings and coverings that representative figures would be meaningless. See Fig. 8.5, and the References quoted

‡ These are the ordinary ceiling tiles, with closed pores, intended for heat insulation and decoration

References

ALKIN, E. G. M., POTTINGER, R. F. A., SALTER, L., and GILFORD, C. L. S. (1962): 'The broadcasting of music in television', *BBC Eng.*, (40), Feb.

ANDRADE, E. N. DA C. (1932): 'The Salle Pleyel', *Nature*, **130**, (3279), pp. 332–333

ATAL, B. S. (1959): 'A semi-empirical method of calculating reverberation chamber coefficients from impedance values', *Acustica*, **9**, pp. 27–30

AXON, P. E., GILFORD, C. L. S., and SHORTER, D. E. L. (1955): 'Artificial reverberation', *Proc. IEE*, **102** B, pp. 624–642

BACKUS, J. (1971): 'The acoustical foundations of music' (John Murray), p. 158

BERANEK, L. L. (1950): 'Acoustic measurements' (Wiley)

BERANEK, L. L. (1953): 'Revised criteria for noise in buildings', *Noise Control*, **3**, pp. 19–27

BERANEK, L. L. (1957): 'Criteria for noise in buildings', *Noise Control*, **3**, pp. 19–27

BERANEK, L. L. (1962): 'Music, acoustics and architecture' (Wiley)

BERANEK, L. L., (Ed.) (1960): 'Noise reduction' (McGraw–Hill), chap. 21, pp. 541–570

BOLT, R. H. (1938–39): 'Frequency distribution of eigentones in a three dimensional continuum', *ibid.*, **10**, pp. 228–234

BREBECK, D., BÜCKLEIN, R., KRAUTH, E., and SPANDÖCK, F. (1967): 'Akustische ähnliche Modelle als Hilfsmittel für die Raumakustik', *Acustica*, **18**, pp. 213–226

BROWN, S. (1964): 'Acoustic design of broadcasting studios', *J. Sound. & Vib.*, **1**, pp. 239–257

BROWN, S. (1965): 'Recording studios for popular music'. Proceedings of the 5th international congress on acoustics, Liège, Paper G36

BRUEL, P. V. (1951): 'Sound insulation and room acoustics' (Chapman & Hall), p. 240

BRYAN, M. E., and PARBROOK, H. D. (1960): 'Just audible thresholds for harmonic distortion', *Acustica*, **10**, pp. 87–91

BS2750: 1956. 'Recommendations for field and laboratory measurement of airborne and impact sound transmission in buildings'

BS3489: 1962. 'Sound level meters, industrial grade'

BS3638: 1963. 'Method of measurement of sound absorption coefficients. ISO'

BS4196: 1967. 'Guide to the selection of methods of measuring noise emitted by machinery'

BS4197 1967. 'A precision sound level meter'

BS4198: 1967. 'Method for calculating loudness'

BURD, A. N. (1958): 'Correlation, first report'. BBC Research Department Report B066 [1958/8]

BURD, A. N. (1964): Correlation techniques in studio testing'. BBC Research Department Report B082 [1964/34]

BURD, A. N. (1968a): 'The measurement of sound insulation in the presence of flanking paths', *J. Sound & Vib.*, **7**, pp. 13–26

BURD, A. N. (1968b): 'The reflection of sound from television scenery flats'. BBC Research Department Report PH24 [1968/42]

BURD, A. N. (1969): 'Non-reverberant music for acoustic model studies'. Report of the EBU conference on the use of scale models for the study of studios and concert halls, Hamburg, Tech3089E, pp. 20–23

BURD, A. N., and GILFORD, C. L. S. (1958): 'Improvements in and relating to sound absorbers', British Patent 860 682

BURD, A. N., GILFORD, C. L. S., and SPRING, N. F. (1966): 'Data for the acoustic design of studios', *BBC Eng.*, (64), Nov.

BURNS, W. (1962): 'Hearing and the ear'. Proceedings of the 12th National Physical Laboratory symposium on the control of noise, Teddington, pp. 19–37 (HMSO)

CHERRY, C. (1963): 'Stereophonic listening—facts and fancies', *J. IEE*, **9**, pp. 66–69

COLEMAN, P. D. (1962): 'Failure to locate the source distance of an unfamiliar sound', *J. Acoust. Soc. Am.*, **34**, pp. 345–346

COOK, R. K., WATERHOUSE, R. V., BERENDT, R. D., EDELMAN, S., and THOMPSON, M. C. (1955): 'Measurement of correlation coefficients in reverberant sound fields', *ibid.*, **27**, pp. 1072–1077

CRANDALL, I. B. (1926): 'Vibrating systems and sound' (van Nostrand), p. 146

CREMER, L. (1942): 'Theorie der Schalldämmung dünner Wände bei Schrägen Einfall', *Akustische Z.*, **7**, pp. 81–104

CREMER, L. (1953): 'Calculation of sound propagation in structures', *Acustica*, **3**, pp. 317–325

CREMER, L., and GILG, J. (1970): 'Zur Problematik der prufgerechten Körperschall Anregung von Decken', *ibid.*, **23**, pp. 54–63

CREMER, L., and HECKL, M. (1967): 'Körperschall' (Springer)

CROCKER, M. J., and PRICE, A. J. (1969): 'Sound transmission using statistical energy analysis', *J. Sound & Vib.*, **9**, pp. 469–486

D˜MMIG, P. (1957): 'Zur Messung der Diffusität von Schallfelder durch Korrelation', *Acustica*, **7**, pp. 388–398

DATE, H., and TOZUKA, Y. (1966): 'An artificial reverberator whose amplitude and reverberation time/frequency characteristic can be controlled independently', *ibid.*, **17**, pp. 42–47

DUBOUT, P. (1958): 'Perception of artificial echoes of medium delay', *ibid.*, **8**, pp. 371–378

DUBOUT, P., and DAVERN, E. (1959): 'Calculation of the statistical absorption coefficient from acoustic tube measurements', *ibid.*, **19**, pp. 15–16

EVANS, E. J., and BAZLEY, E. N. (1956): 'The absorption of sound in air at audio frequencies', *ibid.*, **6**, pp. 238–245

EVANS, E. J., and BAZLEY, E. N. (1960): 'Sound absorbing materials' (HMSO)

EYRING, C. F. (1930): 'Reverberation time in dead rooms', *J. Acoust. Soc. Am.*, **1**, pp. 217–241

FLUGRATH, J. M. (1969): 'Modern day rock and roll music and damage risk criteria', *ibid.*, **45**, pp. 704–711

FORD, R. D., LORD, P., and WALKER, A. W. (1967): 'Sound transmission through sandwich constructions', *J. Sound & Vib.*, **5**, pp. 9–21

FORD, R. D., LORD, P., and WILLIAMS, P. C. (1967): 'The influence of absorbent linings on the transmission loss of double-leaf partitions', *ibid.*, **5**, pp. 22–28

FURDUEV, V. (1965): Évaluation objective de l'acoustique des salles'. Proceedings of the 5th international congress on acoustics, Liège, Vol. 1, pp. 41–54

FURDUEV, V., and CH'ENG T'UNG (1960): 'Measurement of the diffuseness of the acoustic field in rooms by the directional microphone method', *Sov. Phys. Acoust.*, **6**, pp. 103–111

GALAMBOS, R., and DAVIS, H. (1943): 'The response of single nerve fibres to acoustic stimulation', *J. Nuerophysiology*, **6**, pp. 39–58

GEDDES, W. K. E. (1968): 'The assessment of noise in audio-frequency circuits'. BBC Research Department Report EL17 [1968/8]

GEDDES, W. K. E., and GILFORD, C. L. S. (1957): 'A new test signal for acoustic measurements: The coherent constant percentage bandwidth pulse'. BBC Research Department Report B065 [1957/3]

GERSHMAN, S. (1951): 'The correlation coefficient as a criterion of the acoustical quality of a closed room', *Zh. Tech. Phys. (USSR)*, **21**, pp. 1492–1496

GILFORD, C. L. S. (1952): 'Helmholtz resonators as acoustic treatment of broadcasting studios', *Br. J. Appl. Phys.*, **3**, pp. 86–92

GILFORD, C. L. S. (1952–53): 'Membrane sound absorbers and their application to broadcasting studios', *BBC Q.*, **7**, pp. 246–256

GILFORD, C. L. S. (1959): 'The acoustic design of talks studios and listening rooms', *Proc. IEE*, **106** B, pp. 245–258

GILFORD, C. L. S. (1967): 'Background noise in broadcasting studios', *IEE Con. Publ.* 26, pp. 91–94

GILFORD, C. L. S., and DRUCE, N. C. H. (1959): 'Wide band absorbers with impermeable facings'. Proceedings of the 3rd international congress on acoustics, Stuttgart, Vol. 2, pp. 853–857 (Elsevier)

GILFORD, C. L. S., and GIBBS, B. M. (1971): 'Internal damping as a factor of sound propagation in building structures'. Proceedings of the 7th international congress on acoustics, Budapest, Vol. 2, pp. 85–88

GILFORD, C. L. S., and GREENWAY, M. W. (1956): 'The application of phase-coherent methods and correlation to room acoustics', *BBC Eng.*, (9), pp. 5–14

GILFORD, C. L. S., and JONES, D. K. (1967): 'Psychoacoustic criteria for background noise and sound insulation in broadcasting studios'. BBC Research Department Report PH5 [1967/11]

GILFORD, C. L. S., JONES, D. K., and MOFFAT, M. E. B. (1966): 'The Television Centre Ambiophony system'. BBC Research Department Report. B088 [1960/2]

GILFORD, C. L. S., and SOMERVILLE, T. (1950): 'Discrimination of pitch in short pulses of sound', *Nature*, **165**, (4199), p. 643

GOFF, K. W. (1955): 'An analogue electronic correlator for acoustic measurements', *J. Acoust. Soc. Am.*, **27**, pp. 223–246

GÖSELE, K. (1953): 'Schallabstrahlung von Platten', *Acustica*, **3**, pp. 243–248

GÖSELE, K. (1954): 'Der Einfluss der Hauskonstruction auf die Schallslangsleitung bei Bauten', Veröffentlichen aus dem Institut für technische Physik, Stuttgart, (34), pp. 282–290

GRACEY, W. (1968): 'Practical noise control at international airports,' *Appl. Acoust.*, **1**, pp. 241–255

HAAS, H. (1951): 'Uber den Einflüss eines Einfachechos auf die Hörsamkeit von Sprache', *Acustica*, **1**, pp. 49–58

HARWOOD, H. D. (1970): 'The use of a model as a diagnostic tool', British Acoustical Society Paper 70/112

HARWOOD, H. D., and GILFORD, C. L. S. (1969): 'Aspects of high quality monitoring loudspeakers. Pt. 2—Monitoring loudspeaker quality in television sound control rooms', *BBC Eng.*, (78), Sept., pp. 10–16

HEAD, J. W. (1953): 'The effect of wall shape on the decay of sound in an enclosure', *Acustica*, **3**, pp. 174–180

HEAD, J. W. (1965): 'The design of gradual transition (wedge) absorbers for a free-field room', *Br. J. Appl. Phys.*, **16**, pp. 1009–1014

INGÅRD, U., and BOLT, R. H. (1951): 'Absorption coefficient of acoustic materials with perforated facings', *J. Acoust. Soc. Am.*, **23**, pp. 533–540

JONES, D. K. (1967a): 'A subjective investigation into preferred microphone balances', *BBC Eng.*, (68), July

JONES, D. K. (1967b): 'The use of flexible couplings to reduce the transmission of vibrations through water pipes'. BBC Research Department Report PH6 [1967/18]

JONES, D. K. (1967c): 'The subjective significance of secondary reverberation'. BBC Research Department Report PH14

KNUDSEN, V. O. (1923): 'The sensibility of the ear to small differences in intensity and frequency', *Phys. Rev.*, **21**, pp. 84–102

KOSTEN, C. W. (1949): 'Note on similarity tests of sound insulation'. Report of the 1948 Summer symposium of the Physical Society Acoustics Group, London, pp. 87–89

KOSTEN, C. W. (1959): 'Die Messung der Schallabsorption von Materialen'. Proceedings of the 3rd international congress on acoustics, Stuttgart, Vol. 2, pp. 815–830 (Elsevier)

KOSTEN, C. W. (1960): 'The mean free path in room acoustics', *Acustica*, **10**, pp. 245–250

KOSTEN, C. W. (1965–66): 'New method for the calculation of the reverberation time of halls for public assembly', *ibid.*, **17**, pp. 325–330

KOSTEN, C. W., and VAN OS, G. J. (1962): 'Community reaction criteria for external noise'. Proceedings of the 12th National Physical Laboratory symposium on the control of noise, Teddington, pp. 373–379 (HMSO)

KOYASU, M. (1958): 'On the relation between the reverberant absorption coefficient and the normal incidence coefficient of fibrous materials', *J. Acoust. Soc. Am.*, **30**, pp. 1163–1164

KRAEMER, F. W., LICHTENHAN, G., and OESTERLIN, D. (1963): 'Der grosse Sendersaal des NDR in Hannover', *Rundfunktech. Mitt.*, **7**, pp. 262–269

KRYTER, K. D. (1959): 'Scaling human reactions to the sound of aircraft', *J. Acoust. Soc. Am.*, **31**, pp. 1415–1439

KUHL, W. (1954): 'Über Versuche zur Ermittlung der günstigen Nachhallzeit grosser Musikstudios', *Acustica*, 4, pp. 618–634

KUHL, W. (1958): 'The acoustical and technical properties of the reverberation plate', *EBU Rev. A*, (49), pp. 8–14

KUHL, W. (1960): 'Der Einfluss der Kanten auf die Schallabsorption pöroser Materialen', *Acustica*, 10, pp. 264–276

KUHL, W. (1964): 'Zulässige Geräuschpegel in Studios, Konzertsälen und Theatern', *ibid.*, 14, pp. 355–360

KUHL, W. (1965): 'Das Zusammenwirken der direktem Schall, ersten Reflexionen und Nachhall bei der Hörsamkeit von Räumen und bei Schallaufnahmen', *Rundfunktech. Mitt.*, 9, pp. 170–183

KUHL, W., and KAISER, H. (1952): 'Absorption of structure-borne sound in building materials without and with sand-filled cavities', *Acustica*, 2, pp. 179–188

KUHL, W., and KATH, U. (1963): 'Akustische Anforderungen an ein Konzertstudio und ihre Realisierung beim grosser Sendesaal des NDR in Hannover', *Rundfunktech. Mitt.*, 7, pp. 270–277

KUTTRUFF, H. (1963): 'Raumakustsche Korrelationsmessungen mit einfachen Mitteln', *Acustica*, 13, pp. 120–122

KUTTRUFF, H., and JUSOFIE, M. J. (1967–68): 'Nachhallmessungen nach dem Verfahren der intergrierten Impulsantwort', *ibid.*, 19, pp. 56–58

LANSDOWNE, K. F. L. (1964): 'A logarithmic amplifier using non-linear silicon diodes'. BBC Research Department Report B056/2 [1964/52]

LYON, R. H. (1964): 'An energy method for the prediction of noise and vibration transmission'. Proceedings of the 33rd symposium of the US Department of Defense on shock and vibration, pt. 2

LYON, R. H., and EICHLER, E. (1964): 'Random vibration of connected structures', *J. Acoust. Soc. Am.*, 36, pp. 1344–1354

MACKENZIE, R. B. (1970): 'Fifth-scale studies of sound insulation using simulated building elements', British Acoustical Society Paper 70/116

MANKOVSKY, V. S. (1971): 'Acoustics of studios and auditoria' (Focal Press)

MARSHALL, A. H. (1967): 'A note on the importance of cross-section in concert halls', *J. Sound & Vib.*, 5, pp. 100–112

MAXFIELD, J. P., and ALBERSHEIM, W. J. (1947): 'An acoustic constant of closed spaces connected with their apparent liveness', *J. Acoust. Soc. Am.*, 19, pp. 71–79

MAYO, C. G. (1952): 'Standing wave patterns in studio acoustics', *Acustica*, 2, 2, pp. 49–64

MEINEMA, H. E., and JOHNSON, H. A. (1961): 'A new reverberation device for high fidelity systems', *J. Audio Eng. Soc.*, 9, pp. 284–289, 324–326

MEYER, E. (1956): 'Messungen zur Körperschallübertragung an Hand von Modellen', *Acustica*, 6, pp. 51–58

MEYER, E., and KUTTRUFF, H. (1958): 'Akustische Modellversuche zum Aufbau eines Hallraumes', *Nachr. Akad. Wiss. Göttingen 2a*, 6, pp. 99–113

MEYER, E., PARKIN, P. H., OBERST, H., and PURKIS, H. J. (1951): 'A tentative method for the measurement of indirect sound transmission in buildings', *Acustica*, 1, pp. 17–28

MEYER, E., and THIELE, R. (1956): 'Raumakustische Untersuchungen in zahlreichen Konzertsälen und Rundfunkstudios unter Anwendung neuerer Messverfahren', *ibid.*, 6, pp. 425–444

MILLS, A. W. (1958): 'The minimum audible angle', *J. Acoust. Soc. Am.*, **30**, pp. 237–246

MINTZ, F., and TYZZER, F. G. (1952): 'A loudness chart for octave band data on complex sounds', *ibid.*, **24**, p. 80

MOFFAT, M. E. B. (1967): 'An analogue to digital converter and paper tape transcoder for processing acoustic signals'. BBC Research Department Report PH8 [1967/29]

MOFFAT, M. E. B., and SPRING, N. F. (1969): 'An automatic method for the measurement of reverberation time', *BBC Eng.*, (80)

MORRIS, D. (1971): 'The human zoo' (Corgi), p. 97

MULHOLLAND, K. A., PARBROOK, H. D., and CUMMINGS, A. (1967): 'The transmission loss of double panels', *J. Sound & Vib.*, **6**, pp. 324–334

MUNCEY, R. W., NICKSON, A. F. B., and DUBOUT, P. (1953): 'The acceptability of speech and music with a single artificial echo', *Acustica*, **3**, pp. 168–173

NIESE, H. (1961): 'Die messtechnische Ermittlung der Lautstärke von beliebigen Geräusche unter Berücksichtigung des Zeitablaufes und der Spektrumsform', *ibid.*, **11**, pp. 70–80

NIESE, H. (1956): 'Eine Methode zur Bestimmung der Lautstärke beliebige Geräusche', *ibid.*, **15**, pp. 117–126

NIMURA, T., and SHIBYAMA, K. (1957): 'Effect of the splayed walls of a room on the sound transmission characteristics', *J. Acoust. Soc. Am.*, **29**, pp. 85–93

OLSON, H. F., and MASSA, F. (1943): 'Dynamical analogies' (van Nostrand)

PARKIN, P. H., and MORGAN, K. (1970): ' "Assisted resonance" in the Royal Festival Hall', *J. Acoust. Soc. Am.*, **48**, pp. 1025–1035

PEDERSON, P. O. (1940): 'Lydtekniske Undersølgelser', *Ingeniorvidensk Skr.* (5), p. 194

PETERSON, G. E., and BARNEY, H. A. (1952): 'Control methods used in a study of the vowels', *J. Acoust. Soc. Am.*, **24**, pp. 175–184

RAES, A. C. (1954): 'Tentative method for the measurement of sound transmission losses in unfinished buildings', *ibid.*, **27**, pp. 98–102

RANDALL, K. E., and WARD, F. L. (1960): 'Diffusion of sound in small rooms', *Proc. IEE*, **107** B, pp. 439–450

RAYLEIGH, LORD (1876): 'Our perception of the direction of a source of sound', *Nature*, **14**, (1), pp. 32–33

RAYLEIGH, LORD (1896): 'The theory of sound—Vol. 2' (Macmillan), p. 70

RIESZ, R. R. (1928): 'Differential sensitivity of the ear to pure tone stimulation', *Phys. Rev.*, **31**, pp. 867–875

ROBINSON, D. W., COPELAND, W. C., and RENNIE, A. J. (1961): 'Motor vehicle noise measurement', *Engineer*, **211**, (5488), pp. 482–497

ROBINSON, D. W., and WHITTLE, L. S. (1964): 'The loudness of octave bands of noise', *Acustica*, **14**, pp. 24–35

ROBINSON, D. W., WHITTLE, L. S., and BOWSHER, J. M. (1961): 'The loudness of diffuse sound fields', *ibid.*, **11**, pp. 397–404

ROFFLER, S. K., and BUTLER, R. A. (1968): 'Factors that influence the localisation of sound in a vertical plane', *J. Acoust. Soc. Am.*, **43**, pp. 1255–1259

RSCHEVKIN, S. N. (1938): 'Resonance absorption of sound', *C.R. Akad. Sci.*, **18**, pp. 564–576

SABINE, W. C. (1922): 'Collected papers on acoustics' (Harvard University Press)

SAYERS, B. M., and CHERRY, E. C. (1957): 'The mechanism of binaural fusion', *J. Acoust. Soc. Am.*, **29**, pp. 973–987

SCHROEDER, M. R. (1954): 'Statistik der Frequenzkurve in Räumen', *Acustica*, **4**, pp. 594–600

SCHROEDER, M. R. (1959): 'Measurement of reverberation time by counting phase coincidences'. Proceedings of the 3rd international congress on acoustics, Stuttgart, Vol. 2, pp. 597–905

SCHROEDER, M. R. (1965): 'New method of measuring reverberation time', *J. Acoust. Soc. Am.*, **37**, pp. 409–412

SCHROEDER, M. R., and LOGAN, B. F. (1961): 'Colorless artificial reverberation', *J. Audio Eng. Soc.*, **9**, pp. 192–197

SERAPHIM, H. P. (1958): 'Untersuchungen über die Unterschiedschwelle exponentiellen Abklingens von Rauschbandimpulsen', *Acustica*, **8**, pp. 280–284

SMITH, T. J. B., and GILFORD, C. L. S. (1968): 'Airborne sound insulation requirements for studio centres'. BBC Research Department Report PH21 [1968/35]

SNOWDON, J. C. (1968): 'Vibration and shock in damped mechanical systems' (Wiley)

SOBOLEV, A. (1955): 'A new type of ball-valve', *J. Plumbing Trades Un.*, **3**, pp. 234–236

SOMERVILLE, T. (1953): 'An empirical acoustic criterion', *Acustica*, **3**, pp. 365–369

SOMERVILLE, T., and BROWNLESS, S. F. (1949): 'Listeners' sound level preferences', *BBC Q.*, **3**, pp. 12–26

SOMERVILLE, T., and GILFORD, C. L. S. (1952): 'Cathode-ray tube displays of acoustic phenomena and their interpretation', *ibid.*, **8**, pp. 41–53

SOMERVILLE, T., and GILFORD, C. L. S. (1957): 'Acoustics of large orchestral studios and concert halls', *Proc. IEE*, **104**B, pp. 85–97

SOMERVILLE, T., GILFORD, C. L. S., NEGUS, R. D. M., and SPRING, N. F. (1966): 'Recent work on the effects of reflectors in concert halls and music studios', *J. Sound & Vib.*, **3**, pp. 127–134

SOMERVILLE, T., and HEAD, J. W. (1957): 'Empirical Acoustic criterion', *Acustica*, **7**, pp. 96–100

SOMERVILLE, T., and WARD, F. L. (1951): 'Investigation of sound diffusion in rooms by means of a model', *ibid.*, **1**, pp. 40–48

SPANDÖCK, F. (1934): 'Akustische Modellversuche', *Ann. Phys. (Germany)*, **20**, pp. 345–360

SPANDÖCK, F. (1964): 'Die Vorausbestimmung der Akustik eines Raumes mit Hilfe von Modellversuchen'. Proceedings of the 5th international congress on acoustics, Liège, Vol. 1, pp. 313–344

SPRING, N. F. (1971): 'Automatic reverberation time measurements with the aid of a computer'. Proceedings of the international congress on acoustics, Budapest, Vol. 2, pp. 201–204

SPRING, N. F., and RANDALL, K. E. (1969): 'The measurement of the sound diffusion index in rooms'. BBC Research Department Report [1969/16]

SPRING, N. F., and RANDALL, K. E. (1970): 'Permissible bass rise in talks studios', *BBC Eng.*, (83), pp. 29–34

STEVENS, S. S. (1956): 'Calculation of the loudness of a complex noise', *J. Acoust. Soc. Am.*, **28**, pp. 807–832

STEVENS, S. S., WARSHOFSKY, F., and editors of *Life* (1965): 'Sound and hearing' (Time–Life International), p. 32

STRUVE, W. (1964): 'Bau- und Raumakustik im Studio-Neubau des Bayerischen Rundfunks', *Rundfunktech. Mitt.*, **8**, pp. 10–18

TARNOCZY, T. (1958): 'Uber der vorwärts-rückwarts Eindruck', *Acustica*, **8**, p. 343

THIELE, R. (1953): 'Richtungsverteilung und Zeitfolge der Schall-rückwurfe in Raume', *ibid*, **3**, pp. 291–302

THURLOW, W. R., MANGELS, J. W., and RUNGE, P. S. (1967): 'Head movements during sound localisation', *J. Acoust. Soc. Am.*, **42**, pp. 489–493

VAN LEEUVEN, F. J. (1960): 'The damping of eigentones in small rooms', *EBU Rev. A*, **62**, pp. 155–161

VAN LEEUVEN, F. J. (1965): 'A miniature reverberation plate'. Proceedings of the 5th international congress on acoustics, Liège, Paper H68

VERMEULEN, R. (1958): 'Stereo-reverberation', *J. Audio Eng. Soc.*, **6**, pp. 124–130

VON BEKESY, G. (1955): 'Human skin perception of travelling waves similar to those on the cochlea', *J. Acoust. Soc. Am.*, **27**, pp. 830–841

VENZKE, G. (1954): 'Zur Raum- und Bauakustik der Mehrzweckstudios im Hamburger Funkhaus', *Tech. Hausmitteilungen NWDR*, **6**, pp. 229–236

WALLER, R. A. (1966): 'Building on springs', *Nature*, **211**, (5051), pp. 794–796

WARD, F. L. (1952): 'Helmholtz resonators as acoustic treatment in the new Swansea studios', *BBC Q.*, **7**, pp. 7–14

WARD, F. L., (1962): 'The accelerometer as a diagnostic tool in noise studies in buildings', Proceedings of the 4th international congress on acoustics, Copenhagen, Paper L11

WARD, F. L., and RANDALL, K. E. (1966): 'Investigation of the sound insulation of concrete slab floors', *J. Sound & Vib.*, **3**, pp. 205–215

WATERHOUSE, R. V. (1955): 'Interference patterns in reverberant sound fields', *J. Acoust. Soc. Am.*, **27**, pp. 247–252

WATTERS, B. G. (1959): 'Transmission loss of some masonry walls', *ibid.*, **31**, pp. 898–911

WESTPHAL, W. (1954): 'Zur Schallabstrahlung von Wanden', *Acustica*, **4**, pp. 603–610

WHITFIELD, I. C. (1967): 'The auditory pathways' (Arnold)

WÖHLE, W. (1956): 'Zum Schallpegel in Ecken, Kanten und an den Wänden geschlossenen räumer bei Rauschen', *Hochfrequenztech. & Elektroakust.*, **64**, pp. 158–161

WÖHLE, W. (1959): 'Die Schallabsorption von Einzelresonatoren in verschiedenen Anordnungen im geschlossenen Raum. (Wandmitte, Kante, Ecken)', *ibid.*, **67**, pp. 80–97

ZWICKER, E. (1958): 'Über physiologische und methodologische Grundlagen der Lautheit', *Acustica*, **8**, pp. 237–258

ZWICKER, E. (1960): 'Ein Verfahren zur Berechnung der Lautstärke', *ibid.*, **10**, pp. 304–308

ZWICKER, E. (1967): 'Lautstärkeberechnungsverfahren im Vergleich', *ibid.*, **17**, pp. 278–284

Index